Is The Truth Out There?

Is The Truth Out There?

◆

A Journey Through Critical Thinking that Spans Man's History, Origin and Place in the Universe

Darrick Dean

iUniverse, Inc.
New York Lincoln Shanghai

Is The Truth Out There?

A Journey Through Critical Thinking that Spans Man's History, Origin and Place in the Universe

iUniverse, Inc.

For information address:
iUniverse, Inc.
2021 Pine Lake Road, Suite 100
Lincoln, NE 68512
www.iuniverse.com

ISBN: 0-595-29185-6

Printed in the United States of America

Wisdom calls aloud in the street.
she raises her voice in the public squares;
at the head of the noisy streets she cries out,
in the gateways of the city she makes her speech…

Blessed is the man who finds wisdom,
the man who gains understanding,
for she is more profitable than silver
and yields better returns than gold…

…preserve sound judgment and discernment,
do not let them out of your sight;
they will be life for you…

Hold on to instruction, do not let it go;
guard it well, for it is your life…

"…whoever fails to find me [wisdom] harms himself;
all who hate me love death."

—King Solomon

Contents

Acknowledgments

At one level, this book is a review or summary of subjects I have examined and studied over a number of years. It also serves to connect these subjects through the eyes of critical thought and find the relations—or the overarching "big picture"—between these common issues. Finally, this book may help to clear my cluttered mind. Then again, that may take a couple more books.

To assist in accomplishing this, I assembled a "review team" to insure accuracy, readability and some minimal level of coherence. Their backgrounds range from one end of the spectrum to another: Physics to theology, business to engineering, biology to education, chemistry to accounting and so on. As we will see, those fields are not so distant from each other and most of them came into play in this book. Maybe the reviewers would not have written or approached this project in the same style or methods as I, or agree on every last point, but their reviews and suggestions have made this a much more understandable and accurate book.

Thanks to modern technology, this team was able to participate from far-flung locales on two continents. I would like to thank Kyle Witten, Alejandro Field, Bonnie O'Neil, Richard Deem, Mark P. Perez, Ralph Maurer, Greg Moore, Ed Mahlum, Carl Shannon, Matt Malingowski and Rich Dean for reviewing parts of the book or its entirety. Parts of this book also benefited from discussions with people on these topics through various avenues on the Internet over the past few years. These people ran the gamut in backgrounds and beliefs—both for and against what is written in this book—but this book gained refinement from all of them. In spite of the Internet's reputation for containing a great deal of fallacious information, it still serves as an ideal place for reviewing, refining and questioning ideas and theories.

Lastly, but certainly not least, I would like to thank my wife for reviewing this book in a brave attempt to understand me better, as well as suggesting many useful clarifications.

Introduction

Reason is the human ability to determine what is real or not real by *thinking*.—Phillip E. Johnson[1]

"Thinking? I didn't know this book was going to make me think! I thought this book was going to help me make sense of everything!?"

Do not panic! Thinking critically is not hard. You will not need to memorize the equations of special relativity or the intricacies of a black hole's structure. In fact, making sense of this world starts with a simple process. Yes, it may take more than thirty seconds of your time, but reasoning through a problem or issue is not inherently difficult. However, it is inherently necessary. And perhaps, just perhaps, it will let you make sense out of a thing or two.

Take a look at your newspaper, the television programs you watch or the books you read. How do you know what they say is true? What is real and what is fraudulent? Are those conclusions by authorities as concrete as they make them sound? Which arguments are hiding illogical or irrational ideas? This may seem to be an impossible problem because none of us can know everything. At the very least, though, we can know some basics to help us understand the world around us in order to make decisions, to understand issues and to vote. We also do this to instill in the next generation ideas, thoughts and a consistent worldview that improves—and perhaps preserves—their existence on this planet.

How many people do you know that believe something is true, yet cannot explain why they accept it as truth? This leads us to the Ultimate Question: *How can we know there is objective truth or an objective basis to morality (moral truth)?* Is this as hard as it sounds? Well, the idea of *truth* presupposes that there are unchanging facts and laws. Are there such things or are there valid truths that differ for each person or belief system? We will attempt to answer that question in Chapter 1.

Education is highly specialized. If you study business, you learn little science. A biologist? Not much physics. An English teacher? Very little mathematics. The system of colleges and universities is supposed to overcome this through liberal arts programs (Definition Clarification: *Liberal* defined as "broad-minded" covering a little of each subject, not liberal in the sense of a political viewpoint). However, with each field becoming more specialized, less and less time remains for others. Is there a set of tools that will help overcome this problem to some extent? Yes there is.

We will examine in Chapter 2 the tools that everyone can learn that will allow them to detect bad argumentation or questionable conclusions. You do not need to be an "expert" in politics, science or whichever field that is in question to separate useful information from the useless. "So you're saying it's an easy, cut and dry process? Just keep this book with me at all times?" No, not exactly. Understanding these tools is only a starting point. The comprehensive principles underlying these tools could be summed up by these Maxims:

1.   Examine both sides of an argument.

2.   Determine where the arguments fail to follow reason or the reasoning process.

3.   Test the evidence the argument presents.

4.   Keep an open, honest mind and allow the facts to conclude what they may.

5.   Start this process over if new, contrary ideas, theories or arguments are raised.

For subjects of passing interest or those you encounter in your daily lives, these maxims and the tools they encompass will help you to determine truth from fiction. Ultimately, if you want to understand any topic in more detail, you must commit time to studying it. This is especially true of subjects in which you specialize or those that hold some special meaning to you. However, and perhaps to your annoyance, the more you study a topic, the quicker you may realize how it is interconnected to many other topics. Yes, that is something else this book will demonstrate.

The basic tools being outlined here will point you in the right direction. Be warned: The tools of clear thinking can be abused! The best way to avoid this pitfall is to return to Maxim #4. A mind committed to honest inquiry avoids going astray. Having said that, keep in mind that none of us are perfect. No one always gets everything right. Nevertheless, we can at least try to be right most of the time.

Other reasons that this guide is needed: Our society has become used to "sound bite" news as opposed to in-depth, detailed debate and discussion. We accept science as being part of our lives (even if we do not understand it) to the point that we forget that those speaking about it are not exempt from making mistakes. The study of history is often brief and watered-down to the extent that the mistakes of the past are repeated. Our government has to spend tax money to teach parents to be responsible for their children's behavior and education (Such as those ads that say talk to your kids and spend time with them. No kidding!). Churches spend little time on the intellectual side of faith that was once so integral to their beliefs. College and university professors teach their personal views and relativistic philosophies instead of hard facts and reasonable science (Sadly, the worst educational institutions are usually the government funded variety).

Those are the types of subjects we will examine in the following pages. This format is a little different than similar books on critical thinking, bad science and related matters. Most of them do not begin by laying out the rules of critical thought. Perhaps some of them are trying to hide their own bias or assume you will accept their authority. Others do discuss critical thought only to abuse the rules in order to convince you of their personal beliefs. At the onset, I am detailing the Maxims and Rules in this introduction and the first two chapters so you can arrive at conclusions for yourself.

This book is not about reviewing or debunking every theory, proposal and oddball claim in great detail. Any one of the topics in this book has many resources devoted to the subject and some are listed in the Notes and Bibliography section of this book. On the other hand, we are going to see that many arguments on which people base their views often fall apart very quickly with simple analysis. The whole point of this book is to provide a concise guide and serve as an example in examining arguments and claims. More details on the subjects that interest you the most are left to those other sources.

The essays and discussions in this book will detail everything from history to intelligent design theory in an attempt to illustrate application of the Maxims, Rules, reason and clear thinking. Alternatively, this venture into critical thought also serves as an exploration of "Man's History, Origin and Place in the Universe."

These are the things that will also explore the foundations of wisdom. Wisdom is more than knowledge or trivia, it requires "sound judgment and discernment." Think of the chapters as case studies or illustrations. Indeed, they cover a wide array of topics, but various threads connect them all. The strongest of these threads (more like a chain) is the fact known as: Truth is Not Relative to the Beholder.

That is where we will now begin.

1

Is There Absolute Truth?

Before we address critical thought we have to determine if an ultimate truth standard exists. To do that we need to examine the philosophies that propose there is not such a standard and the ones that counter that claim.

The philosophy of relativism claims that there is not any absolute truth standard. Here "truth" is usually referring to moral truth, but the following discussions could equally apply to overall objective truth (as in the facts that serve as the basis of history). Relativists claim what is right for you may not be right for someone else. "Do your own thing." "More power to you." These are phrases of our relativistic thinking. We are not talking about simple personal preferences. If you like seafood and your spouse does not, that is not relativism. Relativism is stating there is nothing to determine *absolutely* whose morals or laws are correct. You may already be thinking that this does not make sense. If everything is relative, how can we make laws and enforce them? You are on the right track, but first we must examine some more terms and definitions.

A worldview is the overarching or fundamental belief system, or set of beliefs, that governs all aspects of your life or is the "glasses" through which you perceive the world. It determines how you act, whom you vote for, what you read or the reasons and motivations behind your studying a particular subject. There are a variety of worldviews such as theism, deism, monism and naturalism.

Theism postulates that a transcendent designer created the universe (Big Word "transcendent" Basically Means: To be separate from the universe and not bound by it. Any being independent from the universe would also be capable of interacting with it.). This worldview can be broken down into three main categories. Monotheism conceives of one God (Judaism, Christianity and Islam are monotheistic). Polytheism supports beliefs in many gods (such as the ancient Greeks and Romans). Pantheism sees the world as God and all are divine (New Age movement). Other worldviews include deism (God exists but does not

1

involve himself in the world) and monism (an impersonal "oneness" such as in Hinduism, Buddhism and New Age also appeals to this).

Naturalism claims that everything in this universe boils down to nature. Everything that occurs does so through random processes or seemingly coherent processes (such as the laws of physics) that are somehow created by unguided nature. Chance, or more specifically probability, replaces any outside influence—such as a group of gods, one God or a designer. In effect, the god of naturalism is nature or chance. One may ask, "What is chance? Isn't it just the idea of probability or accident? How can it be talked about as a force or god?" If you have asked these questions, you have already discovered a major problem with naturalism. It speaks about chance and random processes almost as if they have some sort of inherent supernatural power. The problem is that chance is a nonentity and does not have any physical or metaphysical reality. It is a concept and term in statistical mathematics and "games of chance." Any usage beyond that and the naturalists have unknowingly (or unwillingly) created their own all-powerful entity.

But what if chance-based naturalism is true? Then there is no ultimate source of moral truth, hence naturalism is a major source of relativism. In later chapters we see how naturalism can corrupt reasonable science or the interpretation of natural truth. Natural truth refers to the laws of physics, which are accepted as such only when a significant understanding of their precision is achieved. These laws do not change, although our *understanding* of them sometimes changes as our knowledge and technology improves. If these laws were subject to change, chaos would ensue and it is unlikely that the universe would have made it this far. This does not stop those who would like to ignore or reinterpret these laws to prop up their personal beliefs or theories.

First, let us stop and determine if relativism is logically valid. There is a test that has been proposed to examine whether or not a worldview or belief system is true.[1] It consists of these three criteria, which we refer to as our Rules:

1.   Logical Consistency: A true worldview will not contradict itself.

2.   Factuality: A true worldview will fit the facts.

3.   Viability: One can live consistently with a worldview that is true.

How does relativism stand up to this test? Well, it fails the very first criterion.

Remember that relativism states that there is no absolute moral truth or such a thing as objective facts. Read that again. It is stating that there are *absolutely* no absolutes. By making such a statement one is stating what they believe to be an absolute. Yet that absolute claims there are no absolutes. Hence you have a very simple logical contradiction. A claim that contradicts itself cannot be true as for-mulated.

If relativism's premise is untrue, then the converse must be true. Since there cannot be "no absolutes," there must be some underlying absolute truth. Stating that there *are* absolutes does not contradict itself. Therefore logical reasoning reveals there is ultimate or objective truth. This ultimate moral truth would have to be *independent* of the universe—or the universe would need to be shown as being intelligently designed—for it to be meaningful. If morality simply evolved, it is then subject to change. That would make morality and truth (or moral truth) relative and we have seen how this is illogical. Morality suggests a level of com-plexity and intelligence that is not capable of being produced by random pro-cesses. We will discuss in later chapters that nature can produce order (because nature itself is based on order), but not complex specified information such as a coherent, moral "code" or objective standard of truth.

Also realize that relativistic thinkers will downplay the inherent contradiction in their thinking by appealing to the validity of natural laws. However, this too is a contradiction. If there are no absolutes, then even natural law is up for interpre-tation. These thinkers sometimes will pick and choose when science is valid and when it is not, according to what belief they are trying to prove. This selective use of science is not only the realm of some naturalists, but even in the non-relativis-tic worldview of theism some reinterpret science to their own ends.

You may be asking, "If relativism fails so easily, wouldn't its big brother natu-ralism face some significant problems as well?" You are right, it does, but that is left for another chapter. The validity of these philosophies and worldviews will be further tested by answering questions such as: Is mankind really inherently good? Is nature really all that there is or ever was? Can science say nothing about a cre-ator or determine if there was one? Does chance bring about complexity and intelligence? How does theism stand up to the worldview test?

The worldviews of naturalism and theism define our civilization and society more than any other belief systems. The products of naturalism—relativism, atheism and scientific naturalism (or scientism)—and Christian theism are the specific beliefs (or philosophies and religions) that have the most influence on our society (Western Civilization, which in turn has a lot of influence on the rest of the world). They will be at the center of most of our discussions.

Some Things to Keep in Mind: You may see philosophies such as relativism, atheism and scientism classified under the heading of secularism. Some put secularism under naturalism, but they are essentially the same. Here we are referring to belief systems such as relativism as philosophies. Where worship of a deity is involved in the belief system, we refer to it as a religion, such as Christianity. Some refer to beliefs that hold to the lack of a deity, such as in atheism,[2] as religions as well. The fundamental, underlying structure of a belief system—God or nature—is the worldview as is the case in theism and naturalism. Often, instead of saying one has a theistic worldview, one may be more specific and say that one has a Christian worldview, etc. Some may refer to Christianity or atheism specifically as worldviews—and they can be considered such—but this book classifies these things as was just detailed.

Some of the chapters do not focus directly so much on these hefty topics, while in others the link will be more obvious. Nevertheless, your worldview ultimately dictates how you understand, interpret and perceive all of these subjects. Your particular worldview may even make you more susceptible to frauds and misinformation.

In order to determine the validity of worldviews and arguments, we now turn to creating a tool kit for analyzing them.

2

Critical Thinking Tool Kit

o o o o o o o o o o o o o o o o o o o o o o o o o o o o o o o o o
Whenever they are discovered, errors are rooted out, unmasked
and rectified. The proper response to error is not rationalization
but eradication.—William A. Dembski [1]

Reason is best defined as the process of thinking through any idea, subject or
problem. This process uses logic as a tool. When an argument violates logic, it
commits a fallacy. If you want to learn about all of the intricacies and technicali-
ties of logic, buy yourself a textbook on logic. Here we will simply list the major
fallacies that you will encounter in arguments, articles and everywhere else
(Another Definition Clarification: *Argument* refers to the act of discussing a topic
to prove or disprove it, not argument in the sense of yelling and screaming).

Recognition of these fallacies serves as a tool used to test and apply the Maxims and Rules and we could define this tool as the "Fallacies to be Avoided in
Arguing" or "Fallacies to be Recognized in Arguments." For the most part, extensive examples and applications of these fallacies will be left to the following chapters.

Emotionalism (fallacy of diversion or appeal to pity)

This is perhaps the most popular fallacy. Get someone worked up emotionally
and chances are they will be bamboozled into believing you. The technique is relatively simple. Use a lot of "hand-waving" and tear-jerking pleas and people will
think with their emotions instead of their brain. This ploy is very popular in elec-
tions.

Campaign advertisements have been used in an attempt to convince senior citizens that their social security will be taken away. Some even use senior citizens

that sounded scared out of their minds. The truth is that both political parties admitted social security for seniors was and never will be in danger. Yet you would never have guessed that by what was said in those ads. It was simply an emotional plea (or a fabrication hidden by emotion in this example) diverting attention from the authentic issues.

Selective Evidence

Selective evidence is when all contrary evidence to one's belief system or assumed conclusion is ignored regardless of the validity of such evidence. People are reluctant to give up on something they have invested so much time in. One example is the popularity of horoscopes in spite of the evidences that points to their uselessness. Why are all of these "genuine" astrologers not rich and famous for their amazing insights? Supposedly such predictions are based on the movements of the heavens. Go ask your astrologer if he or she knows anything about Kepler's Laws, Einstein's relativity equations or Newton's Law of Gravitation (all describe the movements of the heavens with great precision). It is fairly simple for someone to write predictions so vague that parts of them might apply to nine out of ten people.

In fact, many of these predictions become self-fulfilling. If your horoscope tells you that you will meet someone special when you go out tonight, you may go out instead of sitting at home like you originally planned. Thus your chances of meeting someone just increased. All because you listened to a stranger's vague prediction aimed at no one in particular, rather than them having any insight into your specific future.

Confusing Correlation with Causation, Shortsightedness

We often confuse correlation with causation such as in saying, "The economy is good, so the current president must be responsible." Economics is not that simple, few things are. The economy in particular is affected by laws and trends begun years ago. One needs to look closely at the causes and details.

A close companion of this is ignoring the long-term results and focusing on the short-term (a favorite of the government). For example, the economy may be good now, but without actions such as lowering taxes, investing in research and development and scaling back government bureaucracies, an apparently good economy may have some loose blocks in its foundation.

Begging the Question, Circular Argumentation

Begging the question as in "the sky is blue because its blueness makes it blue" does not prove anything. Such statements are saying "accept this conclusion because this other conclusion is true." It does not actually say why either conclusion is true, just that they are!

Perhaps a clearer example would be, "He is an honest politician, he would never lie to us!" This statement only "begs" the questions: What makes you think he is honest? What evidence exists to prove he did not lie?

Rationalization

People will often replace reason with rationalization. In fact, this is a keystone to most all fallacies. Philosopher Dr. Dallas Willard explains it this way:

> Rationalization is the use of reasoning to make sure that one comes out in the right place. Not long ago the dominant ideal within intellectual circles was to judge the conclusion by the method through which it was derived. If the method was good, you were required to accept the conclusion, at least provisionally. Now, sadly, the method is judged by whether it brings you out at the "right" conclusion, as determined by institutional consensus congealed around glittering personalities. If you don't come out to the "right" conclusion, your method is wrong, and you are probably a bad person. Derisive terminology will be used to describe you.[2]

Fallacies are often used in rationalizing to a preconceived conclusion. These "right conclusions" are decided upon before reasoning and testing actually occurs. This is called deciding *a priori*, or before the fact. *If* one has facts that can support their a priori claim, then it is not a fallacy. But as we will see a number of times in this book, a priori claims are often unsubstantiated and based simply on one's personal opinion or beliefs. True reasoning and science permit a fallacy-free line of argumentation to conclude what it logically leads to, not what someone tells it to decide.

Shifting Definitions, Broad Generalizations

One cannot argue a point, or disprove another, if everyone is not on the same page with how terms are defined. Terms can have separate definitions for differing applications or many meanings depending on context. People will often misuse these definitions or redefine terms to benefit their particular argument. Also, an arguer may use a term that is too broad in order to convince the reader of more than the argument actually allows. At other times the terms used may be over-specialized to avoid larger implications.

Politically Correct Terms

Politically correct terms often try to soften reality. Instead of calling a military action a war, we call it an "operation," "conflict" or "police action." One can argue how a war is legally defined all they want, but just ask the soldiers for their definition. If armies are shooting at each other, and people are dying, those soldiers have no doubt they are at war. Politically correct terms are often used as a tool by people to make their "cause" seem more acceptable over another alternative. For example, some extreme animal rights organizations often make their cause sound pleasant by adding "humane" into their title. This takes attention away from some of their decidedly anti-human polices and beliefs (see Chapter 8).

Appealing to Popular Consensus and to Polls

"Most people believe it, so it must be true" is not a valid proof for an argument. Reality is not always a democracy. Simply because someone sells thousands of books does not mean what they are writing about is true. Ask questions such as: Why do people believe it? Do they have evidence? Can they explain or defend their beliefs? Have they tested their beliefs or do they blindly believe something is true?

Many opinion polls claim to be scientific, but will the results of polling a few hundred or few thousand be the same as when 30 million are polled? Not necessarily, because poll questions can be carefully worded for certain reactions or people may be given little time to think over their answers. This is how bias can be inserted into statistical studies. When properly used, polls and the sampling of populations serve as a valuable tool and are critical to science. However, especially in politics, polls are an often-abused statistical tool.

Appeal to Authority, Quality of Scholarship

Some important person said something is true, so it must be. If only life were that simple. When an "authority" or "expert" speaks on some subject, there are questions that need to be asked:

1. What are their credentials (education, etc.)?

2. Can these credentials be checked?

3. Are the credentials relevant to the subject matter?

4. Do they actually give evidence to support their theory?

5. Do they violate any fallacies?

This is a good starting place, but the points can be abused.

Because someone has a Ph.D. does not necessarily mean that they are an expert in everything or even in their field. It may very well be a good indication that they know a lot about their field (one would hope they would have picked up something in all that education). However, there are plenty of people that are extremely intelligent, having never gone to college or that have only an undergraduate degree. Ultimately, *quality of scholarship* is what determines the expert's authority on a matter and the soundness of their work. The determination of this quality may indicate a particular website's content to be as valid as what is found in a peer reviewed journal. Or it may relegate a journal's article content to the same level as a tabloid article.

One determines the quality of scholarship (the quality of their argument or work) through a number of ways: Does it violate common fallacies (or our Rules and Maxims)? What do the person's peers say about the argument? Also, does it include conclusions that can be tested and checked by you or someone else, or is it merely speculation?

Unprovable Theories, Lack of Testability

Something that cannot be tested is speculation. If it is speculation, it is not science, but is an unscientific hypothesis or theory. A true hypothesis is a claim that can be tested for verification or repudiation. *Theory* can be used in a similar sense but usually should be used to denote an accepted principle. This can be confusing because people often equate theories with something that may or may not be true. In physics, we hear of the theory of relativity (which has nothing to do with relativism). However, this "theory" has accumulated such support through testing that "theory" is somewhat of a misnomer (see Chapter 7). Theories like this are usually referred to as laws. Relativity makes particular predictions that can be tested, which results in either its verification or falsification. That is what makes a hypothesis or theory scientific. If a theory or claim is not falsifiable—meaning it cannot be tested—it is not science. It is wishful thinking.

Hidden Bias

Perhaps other questions that one should ask are "What hidden agenda or bias does the arguer have?" and "Has this irrationally affected their argumentation?" Ultimately one can argue something that supports their beliefs and do so with a high quality of scholarship. However, this is often not the case.

Philosopher Jean-Jacques Rousseau was a major proponent of the idea that man is inherently good and it is society that corrupts him (see Chapter 4). Was he trying to justify his own lifestyle of mistresses and illegitimate children? Did that influence the defending of his theories in spite of their flaws? Keep an eye out for such influences that may be behind arguments. The honest individual will lay out his intentions, motivations or background as a way to make sure the scholarship of the argument is sound.

Straw Man Arguments

This is simply when someone distorts or misrepresents an opposing argument to make their own look better. Basically the opposing view is caricatured or stereotyped in such a way to make the arguer look like the better choice. Or the opposing argument is redefined in such a way to make it easily defeated.

Attack Fallacies

The most popular fallacious argument is the *ad hominem* (attacking the man). This is when the arguer attacks the person instead of their argument. Sometimes this manifests itself in name-calling or similar childish methods. Credibility may be questioned without giving the accused an adequate voice to answer the claims or by ignoring their answers. Ad hominem arguments can be very subtle, casting just a little doubt on the person's reputation without giving them an adequate chance to respond.

Closely related to this fallacy is the appeal to force. This is when someone says, "Listen to me or pay the price" or "I'll beat you up if you don't agree with me." Rarely does it come to this, but the other attack fallacy, ad hominem, is quite common.

Appealing to Ignorance

This is the door to fantasy, which says, "Believe this because it hasn't been disproved." A better approach would be to construct the argument to state the terms of how it would be proved or disproved and what evidences exist to support the proposal. To believe something because it may be vaguely (in one's mind) true is to believe nothing at all. Believing something requires a reason, not a lack of reasons.

A variant on this is, "It hasn't happened yet, but it might." What reason do you have for thinking that? If you do not have a reason, why do you bother believing it?

Is There No End to these Fallacies?

Other fallacies to watch out for are when people fill their statements with totally irrelevant items, called red herrings. These are intended to distract the audience's attention from the real issues. Some of the fallacies already discussed fall into this category. Another popular mistake is oversimplifying a subject such as "there are billions of stars, so there must be billions of beings everywhere" (more on this particular example in Chapter 6).

There are many more variants on these fallacies, but these are the major ones to keep in mind. Many overlap and are related, but all are fairly straightforward. It does not take long to start recognizing the various fallacies that bring down arguments. You probably will be surprised how often you will start spotting such flaws in arguments, the media and from politicians. Having such "tools" is not foolproof. They can be abused or ignored by the same people who claim to follow them.

Theologian Ken Samples summed up the problem of critical versus uncritical thought when he wrote:

> …it's not surprising that people form irrational beliefs. While people sometimes form their beliefs based on rational factors (facts, evidence, credible authority, and so on), at other times they form convictions based on nonra-

tional factors (emotion, self-interest, peer pressure, and so forth). In other words, people sometimes believe what they want to believe.[3]

There will always be people unwilling to change their mind on an issue regardless of the facts amassed against them. Most people, on the other hand, can reason through arguments and research a subject if they let themselves do so. Thinking a subject through in large part depends on your willingness to leave aside your bias and preconceived notions. If you follow the evidence and logical reasoning to where it leads, you will probably find truth more often than not.

A lot of the fallacies discussed above are used by people to prove their beliefs or arguments. This is especially common in attempts to prove ideas that violate the fact that there exists both objective truth and unchanging natural laws. One place where this has become most prevalent is in the study of history.

3

Changing History Without a Time Machine

One would think that the study of history would be fairly straightforward. Record the facts and analyze the events. Historians will tell you that it is not always quite that easy. Often the past is lost to us, so we must speculate, but we speculate based on what we do know. The study of history is bound by the same precepts of critical thought and reason that govern all other fields of inquiry. However, some would rather reinterpret history for their own personal gain or to benefit their social agenda. By *reinterpret* I mean revise or change.

"How on Earth does one change history?" you ask in disbelief. History, by definition, has already happened so it does not change. Some people obviously do not grasp that simple concept and one does not have to look far to find the phenomenon of revisionist history, or deconstructionism.

Shortly after the terrorist attacks on the United States on September 11, 2001, Hollywood began pulling or postponing movies and television programs that dealt with terrorism or any related subject matter. This sensitivity was at first understandable. What was strange is when they began debating whether or not to *remove* images of the World Trade Center from movies yet to be released.

Apparently they thought that the American public would be too upset to see these images. This was the same American public preparing to go to war, with nearly unanimous support, to defend the nation. Hollywood thought they could make people feel better by suppressing reality. Actually they ended up making most people angry. Preview audiences for these movies largely wanted the images of the Twin Towers left in the films.

Instead of not watching movies about terrorism and war, video stores saw an increase in rentals of such films. Even graphic, true stories like the films *Black Hawk Down* and *We Were Soldiers* did very well at theaters among both genders. The nation wanted to see accounts of American soldiers, because their soldiers were now at war facing the same dangers that were being depicted on the big screen.

In a sort of prophetic irony, a few months before September 11th, Hollywood released the film *Pearl Harbor*. Critics were not as happy with this film as they were with the ones that arrived after September 11th. It took a traumatic event to remind them about patriotism. "Flag-waving" films about our history were not politically correct. Old Hollywood never tried to hide the events of World War II and such films were once a very popular subject matter. Granted, many of these films used to focus on the heroism and honor and not the horrors the soldiers went through.

However, they were better than the Hollywood films of recent years that depict war or historic events. Most Vietnam War films focused on politics and isolated unacceptable behavior of some soldiers. Few of the films focused on the bravery and heroism of the majority of troops. In a world where relativism is popular, subjects like life, truth, honor and respect take a backseat to personal issues and causes. No, we should not hide the mistakes and problems of the past. We need to learn from those mistakes. At the same time we should not replace the rest of history with those problems in an attempt to support personal agendas. Sadly, many viewers do not understand the differences between fact and fiction in entertainment.

Consider *Robin Hood: Prince of Thieves* based on the legend of Robin Hood. The historical basis of Robin Hood is murky and debated. In this version, however, even debated historical facts are replaced by fantasy. For example, Robin Hood is a returning prisoner of war from the Crusades who finds his land stolen. For multiculturalism's sake, Robin Hood now has an enlightened, dark-skinned Islamic sidekick without whom Robin Hood and his men could not save the day. It all makes for an exciting movie, but how do such "political corrections" help our perception of the past?

Racism of a New Sort

Under most circumstances, multiculturalism is not a bad idea. That is if it focuses on studying and respecting one's heritage and ancestry. Once infected with relativism, revising history enters the picture to better one's race, group or cause. Multiculturalism then becomes a badge to wear saying, "I'm better than you" or "I deserve this because of my people's past." This type of multiculturalism is a byword for a Twenty-first Century form of racism.

This perverted type of multiculturalism made itself very obvious a few weeks after the September 11th attacks. A model of a proposed monument to the firefighters that died in the attack was unveiled. It was based (sort of) on an actual photo of three firefighters who had raised the American flag over the rubble of the World Trade Center.

In order to be politically correct, the model replaced two of the three *white* firefighters with two of different ethnicities. Many people found this change of reality distasteful. Defenders claim they were only trying to show a cross-section of those who died. Critics responded by asking, "If 'races' are truly equal, why do the colors of the men need changing?" If one is going to base a monument on a photograph seen world over, to change the details does not seem legal. It is also an affront to a recorded moment in history.

It took a long time to purge the "superior white race" mentality from the thinking of many. Instead of preserving this newly found equality some have abused multiculturalism to form new ethnocentrisms. They make other races into "superior races" through the revision of history. This is common with Afrocentrists who revise history, claiming credit for a variety of technological and societal advances. These are extremists by any definition, but they have a lot of influence in some academic circles. All cultures have contributed to civilization, both good and bad. Nevertheless, in spite of its mistakes, Western Civilization has undoubtedly contributed more to *benefit* the modern world then any other. Few honest historians will deny this and would agree that the minimization of "Western Civ" is as wrong as the ignoring of other cultures.

History Untold, Retold

"Touchy-feely" and watered-down education has exposed many students to revised history. Instead of studying the facts as they are, educators, various movements and groups with their own agendas try to reinterpret history through modern eyes or personal beliefs instead of considering the historical facts. Consider the treatment of any war in history books at the elementary and high school level. The accounts are often brief and misleading. One gets the idea that the wars like the Revolution or the World Wars were brief and easy. The suffering and sacrifice that was expended to defend freedoms we take for granted are barely mentioned, if at all. Of course, we cannot write about the graphic detail of war to youngsters, but should we hide the fact that many died to allow them to live free? Should they be left unaware of what it takes to maintain freedom?

Another example of rewriting history occurred in 1992. The 500th anniversary of Columbus' historic voyage to America was celebrated that year. Protesters of the anniversary said Columbus was driven by greed and was a racist responsible for thousands of Indian deaths and was out to conquer the New World. Factual history tells another story.

Columbus' own writings attest that riches or serving the Spanish Empire were not his primary concern. Unlike later explorers, he explored because he felt he was called to do it. He saw himself as a missionary, not a plunderer. Protesters wanted to hold him responsible for the later deaths and exploitation of some natives, with whom Columbus had nothing to do with. Most Europeans did not immigrate to the Americas to conquer, but to live in peace and freedom from oppressive nations.

There was exploitation of many natives, but many Indians actually deserved to be called savages. The Central and South American regions included cannibals and human-sacrificing groups. North America had natives whom were quite brutal in warfare. These peoples were no more pure than anyone else on the planet. The point here is not to single out those with native heritage, but to show that political correctness can give way to the bias of modern social causes. The outcry around Columbus' anniversary shows some groups are bent on punishing the present for the past (the protests are usually repeated every year in some form). Learning from the past and changing our ways is no longer enough for these people. Someone must be held accountable for all crimes in humanity's history even if it means sacrificing the facts surrounding a historical figure.

Causes vs. History

Oddly enough, many in these social causes preach equal treatment or recognition of minority groups, yet "tolerance" seems to apply only to people who agree with them. Even if you accept that all people have the same rights, disagreeing with their lifestyle or beliefs could make *them* intolerant of you. Christians do not condemn homosexuals or deny them their rights, they disagree with the lifestyle these people have chosen. The Christians are then called intolerant at the same time the "social cause" groups are trying to secure "special protections" and "special rights" in the legislatures. Who is being intolerant? Those who demand special rights above others or those who disagree with someone else's lifestyle or beliefs? The United States was founded on the principle of the equality of people *and* the right to disagree, not the elevation of one group over another.

In Texas, some wanted to rewrite textbooks to put less emphasis on the Texas revolution that led to its freedom from Mexico. Since there are many children with Hispanic heritage in Texas, some feel they will be offended by "Remember the Alamo" and Texas' ultimate victory. To change history is degrading to one's heritage, regardless of which side their ancestors were on. How does one learn to ensure a productive future if the past, in all its achievements and mistakes, is unknowable? History should be written to reflect what happened as accurately as possible.

Is flying the Confederate flag racist or is it part of the South's history? Simply because some racists over the years have used the flag as a symbol does not make it a symbol of racism. The Stars and Stripes flew over slavery for decades as well. Should we throw it out too? The revisionists also forget that there were *free* blacks that *chose* to fight on the Confederate side and supported their cause. There were *free* blacks that *owned* slaves. All under the Confederate flag. There is more to history than the blurbs you read in high school. People stamping out racism should not hide the signs of the past, which only dooms us to repeat the same mistakes. Ironically, the anti-flag crowd ultimately hurts its own cause.

There are other examples of abusing history such as the few African Americans who seek payment, or reparations, from the government because their ancestors were slaves. Certain Jews want money from companies that worked for Nazi Germany. Never mind that these companies are not what they once were or even had a choice in some cases. Forget the fact that no one alive today in America ever owned a slave from Africa. Where do we draw the line? Should Christians seek money for being persecuted in the Roman Empire from Italy, where the empire was ruled? Should Israel hold the Iraqi people responsible for when Babylon (which was in modern day Iraq) took their people captive? How about any of the thousands of groups of peoples persecuted by thousands of other groups of peoples throughout the history of mankind? Reparations are becoming a modern form of welfare under the guise of bringing "justice" to crimes against humanity that have long past into history.

Perhaps instead of abusing history and rewriting it, we should try to learn from it. Instead of punishing people alive today for past mistakes they had nothing to do with, why not make sure such events do not happen again?

In the next chapter, we will take a brief look at the interconnected events of recent history. It serves as a case study of how important studying history objectively is to preventing future evils. Evils that can cost many lives and are inherent to man's very nature.

4

The Ghosts of History

As we approached the end of the Twentieth Century, many used that time as a retrospective to review and to predict what is to come. After the most turbulent, yet most productive century in history, what did we learn? In order to avoid the mistakes of the past, ignoring history is never an option. This requires finding a starting point in the past century that defined today's world and touches everyone who lives in it.

At the end of the Nineteenth Century, the world was still largely comprised of empires and colonial subjects. Countries sought power, forged alliances and conquered nations. Centuries of pent up hostilities that resulted from innumerable wars, campaigns and conquests lay simmering beneath the surface. The world was apathetic and did not see the writing on the wall. Someone only needed to light the fire.

When a shot rang out in Sarajevo and an Austrian monarch lay assassinated, who knew that one of the most defining moments in history was about to commence? It was June 1914 and the Great War was about to begin. Largely forgotten now, yet there is no one that the First World War does not touch, alive now or yet to be born.

As the nations picked sides, it was a time when war was still seen as an adventure, a nearly fun one at that. Thousands of Britons and French went off to war, thinking it would all be over by Christmas. They were often the brightest minds of the day. Most never made it home.

America stayed isolated at first. We had been largely cured of "adventurous and fun" war during the Civil War, which introduced the destruction and death of industrialized warfare. Yet our isolation, which was somewhat misguided, would not last. The war had become unprecedented in the number of people, nations and battlefronts it engulfed. The British and French were running out of people and the Russians were pulling out in political turmoil. When Americans started arriving in Europe, with millions more on the way, the Germans began to see the end. They too had been devastated by years of the war and had no hope of reinforcements.

When President Wilson arrived in Europe to take part in deciding the fate of the losers, he was nearly alone in his ideas. He wanted to draw the lines of nations along cultures. Most of the allies wanted to punish the enemies as much as possible and paid little heed to the problems of cultural conflicts and desired independence. Millions had died, yet little had been learned.

Historians consider the period between the World Wars a cease-fire rather than true peacetime. Germany suffered greatly during the post-war period. It was a prime breeding ground for those who hated the victors of World War I. A new movement rose to the head of the government. It was led by a veteran of the defeated German army. His name was Adolf Hitler. When Hitler began his march through Europe, it was not a new war. Old problems had restarted the most destructive conflict in history. The Second was a continuation of the First and the second would be much worse in terms of death and destruction.

The Second World War ended in 1945 but soon begat the Cold War. The communist revolution first laid its foundation in the turmoil of the First World War and it affects the world to this day. Far more have died at the hands of communist leaders than did from Nazi Germany. The end of the Cold War was prematurely declared with collapse of the Soviet Union. China, North Korea and Cuba still dole out oppression and fear under the imaginary utopia communism supposedly produces. Another artifact of the Cold War, the Vietnam War, had a more direct link to the Great War.

Among the many people after the war who pleaded for their people's independence was Ho Chi Minh of Vietnam. His tried to present Woodrow Wilson with a list of grievances against the French rulers of Vietnam, but his request fell on deaf ears. Minh would turn to communism for support and eventually drove the French out. This is the same communist flood that the United States would try to stop. Had America supported the Vietnamese decades earlier, would the Vietnam War ever have been fought?

We helped Afghanistan fend off the Soviets in the 1980s. At one time, Iran and Iraq were our allies in the Cold War. Now, some in those nations support terrorism against us. We returned to Afghanistan in 2001 to destroy terrorism and liberate the nation from a repressive regime. This time we remembered the lessons of World War II: 1. One must destroy or drive dictators and terrorists to the brink of extinction. Only then to they give up in their aggression; and 2. We must help the people we liberate to give them a fruitful future. In March 2003 the second theater of the War on Terror was opened in Iraq. This time the people were freed and the dictator removed.

Still, many do not understand our intent to help rebuild Afghanistan and Iraq after driving out their terrorist-supporting, repressive governments. Our rebuilding of Europe and Japan after World War II created strong allies and people who do not seek to destroy us. We should have done this the first time in Afghanistan. Our leaving of a dictator in Iraq also ignored the past. One has to completely defeat an enemy of the free world or the threat will return (We left their leader in power because the world thought it was a good idea for local stability. Now we wish we had put our own security foremost, instead of listening to other nations). The United Sates is not perfect and has made many mistakes, but it has sacrificed much and spent considerable resources to help the people of the world. However, if history is ignored, we will end up fighting new conflicts and enemies long after the ancient causes are forgotten by most.

The battlefields of Europe are long silent, but they have not disappeared. Monuments and cemeteries trace the line of the Western Front through France, where the Allies fought the Central Powers to a virtual stalemate. Gains were measured in "yards" and this went on for four years, tens of thousands dying for a few yards at a time. Monuments at these battlefields list hundreds of thousands that died there, but often no graves exist since many remains of the soldiers were never found. They were churned into the ground by artillery attacks of an extraordinary scale. This is the war that also introduced submarine and chemical warfare, aerial bombing and tanks: War had never been so devastating.

As we have seen, the ghosts of the Great War have been more far-reaching than artifacts of the land and we have examined only a small percentage of these ghosts. Will we again forget the lessons of history and why these monuments were raised?

Is Man Inherently Good?

Before the First World War began, the intellectual rage was the belief that mankind is inherently good. This quickly lost favor in light of the unprecedented horrors of that war. This belief of inherent goodness is still prevalent among various belief systems. Naturalism is used to claim that humans evolved inherently good and that evil is a product of society or a result of survival of the fittest. This is all a bit of a contradiction since if we did evolve, then tendencies to do evil must have evolved as well. Society would be a product of evolution, so blaming it for evil would be the same as blaming survival of the fittest. It is a no win circular argument used to ignore the reality of the existence of absolute truth that we discussed. Many use their position in academia and society to promote such beliefs, as horrific events in history become more distant. As much as some "intellectuals" try to promote the utopian view of man, something always comes along to prove this view's futility. After the attacks of September 11th, calling the terrorists *evil* was not uncommon. Evil events such as these show that *goodness* is something mankind has to make an effort to achieve.

What is evil? What does it look like? Just look at the nightly news. People who drown their kids, murder their babies or kill their friends. Most look like they could be your neighbors. The face of evil does not have a particular look. Man is inherently drawn to do wrong. We blame situation, education and government, but for every bad example there are examples of those who succeeded having the same problematic backgrounds. Education and the situation one is born into do have an affect on one's life. How parents raise their kids is also an influential factor. However, we are born with the ability to tell right from wrong. Even people with mental disorders have shown the ability to tell the difference. Children in cannibalistic societies do not accept their macabre practice very willingly. Criminals know what they do is wrong and often admit to it. The facts are stacked against the idea that mankind is inherently "good."

Often the government's attempts to fight evil backfire because they do not address the causes. For example, some people would have all guns confiscated, including those owned by law-abiding citizens. The reasoning is that fewer (or no) crimes with guns will be committed. But does this address why the crime was committed in the first place? No, these people would be committing crimes regardless. Murderers can find ways to kill or rob people without guns just as easy as if they had guns. However, criminals still obtain weapons on the black market regardless of laws—laws that often make it hard for law-abiding citizens to buy defensive weapons. In fact, places with the strictest gun laws (such as Britain), is where crime rates have risen: Criminals still have guns and the gunless public are defenseless, making them easy targets. The founders of the United States often talked about an inalienable right to life (Amendment V of the U.S. Constitution states "No person shall be...deprived of life, liberty, or property, without due process of law..."), which is why they protected the means for one to defend themselves (Amendment II states "...the right of the people to keep and bear arms, shall not be infringed."). Also consider that Hitler confiscated weapons from the public so they could not oppose him.

In spite of all this, some still consider the view of man not being inherently good defeatist. In light of history and current events, it is the realistic view of humans. We are capable of great good, but do so against our instincts. The fact is that the utopian dream has killed millions of people. This view specifically claims that man is corrupted by society and its laws, yet contradictorily asserts man can create an utopian society through government. People like Adolf Hitler and the communist leaders of the Soviet Union tried to fashion "ideal" societies. Creating such a society first required them to purge those who did not fit in their vision of an "ideal master plan" for the citizenry. China has the highest execution rate in the world. This is not due to high crime, but due to their attempts to fashion a society that adheres to totalitarian socialism.

No matter how much we advance as a race, there will be someone or some group that practices the vices of evil. People seem to forget this easily. Before September 11th, much of the United States was mired in dangerous apathy. The Soviet Union had collapsed. We began reducing our military. Terrorist attacks were few, so many people largely ignored them. The attacks that did occur were not addressed adequately, to say the least. Then nearly 3000 people were killed in one day. We were abruptly awakened from apathy and forced back into the sometimes-necessary enterprise known as war to defend ourselves.

Even with such events, will the lessons of history be again forgotten or buried in revisionism? Will we remember the American Revolution, Gettysburg, Pearl Harbor, D-Day, Iwo Jima, Hiroshima, Korea, Desert Storm, Mogadishu, September 11th and all other battles and days of infamy? Will we again try to "contain" and "appease" dictators like Hitler or Hussein or ignore terrorists like bin Laden? Will we forget all those who died for freedom and truth only to have more die again? We wish to—and rightfully so—avoid the destructive courses of war, but to ignore evils of the world will only bring us back to the conflicts we wanted to avoid.

As long as we pretend mankind can be cured of evil tendencies through government programs and social reform—or wishfully believe that evil movements of the world can all be cured by "diplomacy" and "peace"—these problems will never go away. History tells us so. The more money we throw at problems the greater these problems become or they persist unchanged. We do not need more money, we need efficiency, standards and the abandonment of relativistic "truths" (which would be the only way to produce effective standards). Then we can return to the basis of a civil modern society: A worldview based on objective truth. Then it will be more difficult for evil tendencies to masquerade as a sickness or consequence of situation or to be passed off as truth by enigmatic personalities. Ignoring or revising history would be propagated by only the foolish few. The archenemy of evil is truth, because the tools of evil are deceit and lies.

Yes, critical thought leads to some serious subjects. Abducting one's mind, however, often starts with much lighter fare. The road of gullibility is paved with aliens, Elvis and many other bizarre potholes.

5

Abducting your Mind

Elvis is alive, somewhere. Maybe he is an alien. There are buildings on the Moon. Mutant people are bred with creatures of all imaginings. This is the kind of "news" that appears in tabloid newspapers and fake news programs. I always thought this "news" was comical at best. Many are obvious fabrications for entertainment. Then I found out there are people who actually believe them.

It seems some people are under the mistaken impression that anything in print or on television must be true or that some law must require such truth. We can only wish it did demand such integrity, but one can print or televise nearly anything they want as long as someone does not consider it libelous. Common sense should tell people that the tabloid variety of "news" or "science" is untrue. However, there has been a resurgence of belief in such frauds in recent years fed by the mass media.

Alien Ideas

Perhaps the biggest of these obsessions is that of "alien visitation." Despite the "vast" number of Americans who believe we have been visited, no proof has ever been presented. The vast majority of sightings can be explained away as natural occurrences, manmade objects, outright frauds or some other reasonable explanations. The next chapter will discuss the probability of alien life existing, but what about all those "true" stories?

Those that have become movies or television shows, such as *Fire in the Sky*, *Alien Autopsy* and *Communion* have been shown to be most likely fraudulent or attributable to causes other than aliens. Can all the many abduction stories be frauds? None have given any incontrovertible evidence. Usually the evidence is conveniently absent or unable to be viewed. Many such stories have been attributed to drug abuse, mental illness, involvement in the occult and simple scams.

Interestingly enough, the information from the aliens often changes. It was once common for these "aliens" to be from Venus, Mars or other planets. Once space probes determined these worlds were hostile to life, the aliens then began (supposedly) saying they were from distant star systems and galaxies!

Often die-hard Unidentified Flying Object (UFO) believers are also involved in the occult. These are the people who see what are classified as "residual UFOs" (RUFOs).[1] These do not seem to have any manmade or natural cause. These RUFOs and the events surrounding them seem peculiarly similar to occurrences related to the occult seen throughout the centuries. Each era describes these RUFOs in terms of their culture and time (or place) in history. In the past they were gods from Mt Olympus, now they are beings from alien worlds. Considering that aliens from other worlds are extremely improbable (as discussed in the next chapter) and most UFOs can be explained away, what are RUFOs?

In today's world where science has confirmed the multidimensional nature of the universe, the supernatural realm can no longer be cast out as myth and superstition. With so many frauds and scammers in the world, it is a difficult avenue to study scientifically. However, some have waded through the muck and concluded that RUFOs may indeed be part of a nonphysical reality. "Nonphysical reality" may sound like a contradiction, but something from beyond our everyday existence (the physical universe bound by the dimensions of height, width, length and time) could be classified as nonphysical by our standards. Appendix A will detail these interesting lines of research further.

Common sense should also tell us a thing or two about the UFO phenomenon. Why would aliens deliver messages to the world through people in backwoods locales? Why would they travel light-years to make circles in crops? The fact is that hoaxers have come forward showing how complex patterns in fields can be made with ropes and boards. Even radiation signatures, which are supposedly a sign of UFOs, can be faked. These hoaxers often play on the vast body of enigmatic and ancient legends of Britain, where most crop circles appear. Such deceptions, however, do not always come in spaceships.

Monsters and Martians

Two long-running myths are the Loch Ness monster and the "face" on Mars. The original and most famous picture of "Nessie," taken in the 1930s was revealed to be a fraud in 1994. This photo had taken its cue from little known Scottish legends and turned hunting for the "monster" into a career (or obsession) for many people for decades. In retrospect, it is not surprising that the monster did not exist, since such a large creature would be easy to find with modern technology. Also consider that there probably would have to be more than one for procreation. In addition, a reptile—which many believed the creature to be—could never survive in the loch's frigid waters.

The two Viking Orbiters imaged Mars in high resolution in the late 1970s. Some images were taken of what appeared to be a large face. In light of Mars' long history of being connected to extraterrestrial life, many thought this was the proof they needed. In the 1990s, Mars Global Surveyor sent back much higher resolution pictures of the "face" only to reveal it is nothing more than a geological formation. Yet some people are still convinced of Martian settlements and government cover-ups. These people also ignore that Mars' environment has never been conducive to intelligent life even in the best of times.

More Tabloid "Journalism"

Perhaps the granddaddy myth of them all is the Bermuda Triangle, but it too lacks factual backing. It seems that nearly 20 percent[2] of all claimed triangle incidents never happened. A similar percent actually happened, but not in the triangle's boundaries, some were thousands of miles away! Most "disappearances" happened on stormy days (imagine that in the home of hurricanes!), not on the clear days as some try to convince us into believing. Statistics show us that the area has no more lost ships than similar places around the world.

A popular theory presented by the tabloid media for years is that the manned Moon landings were a hoax. This gained more popularity recently when one television network presented a program on the subject. Some took it to be "proof," others thought it was entertaining.

Consider these common sense issues: How does a space program, globally employing tens of thousands of people, fake landing on the Moon? Foreign countries, especially our enemy at the time, the Soviet Union, watched every move we made. They made every attempt to beat us to the Moon. Had we not actually gone, they would have known and used it as propaganda to their advantage.

Evidences such as "girders" in Moon photos are held up as proof. These photos can easily be identified as being taken during Earth-bound simulations. Photos taken on the Moon cannot be real, the conspiracy theorists claim, because there are not any stars in the images. This ignores that short exposure times were used on the film (because of the bright Sun) which precluded most (but actually not all) stars from appearing. The Moon "hoax" is similar to the claim the government is hiding aliens and has wrought a vast conspiracy to hide their existence. How does one create such conspiracies that span decades and involve thousands of people? Where is the tangible evidence that such conspiracies exist?

Predicting Nothing

There are some questions to keep in mind when faced with fantastic claims. Is there independent confirmation or are only an isolated few making a claim? Are there easier explanations than time warps or laws of physics which suddenly fail? Has any evidence been suppressed? One perfect example of the latter is that of self-proclaimed prophet Jeanne Dixon.

She usually does not mention that only six of her 72 major predictions came true.[3] The "right" ones were questionable. Some of her predictions included Russia would beat the U.S. to the Moon and World War III would begin 1954. So much for being a prophet.

Joseph Smith, founder of the Mormon church, and the "sleeping prophet" Edgar Cayce are two other prophets whose own predictions proved they did not have any exclusive insight into the future. Chapter nine discusses some of Smith's predictions and revelations. Cayce, who made his predictions while asleep, predicted events such as part of Atlantis was to appear near the Bahamas in the 1960s and records of a lost supercivilization would be found at the Sphinx in Giza in the late 1990s. Much of the world was also supposed to be destroyed by 1998 from geological events of a catastrophic magnitude.[4] We are still waiting.

Remember what we said about astrology? The same concepts apply to would-be prophets. Vague and broad predictions and "psychics" who like to select the hits and ignore the misses. People tend to remember the hits and forget how vague and broad the predictions often were. Some will point to a few individuals in history whose predictions cannot be explained away.[5] However, as we have seen, the fraudulent prophets are easily flushed out.

The Specter of Cancer

Paying attention to details is important in watching the latest "scientific reports" that the media airs or prints. Whenever a report makes a declaration such as "this substance causes cancer!" often the media will headline it without examining it closely. For example, the sugar-substitute saccharin was listed as a possible cancer causing agent years ago. Today it is mainly used only in low-consumption pharmaceuticals, health products such as toothpaste or by people who have to watch their sugar intake.

The problem with these studies (and similar ones considering aspartame, known better as the brand NutraSweet) is that they are based on cancer occurrences in rats that have been pumped with quantities of the chemicals no normal human would ingest. A human has a higher tolerance to such chemicals and can filter them more efficiently. Needless to say, these substances are no longer considered cancer risks.

This has all been superceded by the "cell phone" scare. Some claim that the radiation emitted by the phones can cause brain cancer. This claim is based on exposure to levels of radiation (watts per kilogram, or "specific absorption rate") far greater than phones actually emit. By the standards of these "scientific studies," our televisions, computers and radios should have killed everyone decades ago. "Scientific" studies like this can make it hard to determine what substances actually are harmful.

This does not mean you should go out and start smoking again. There are substances that are well established to be cancer causers, such as ultraviolet radiation from the Sun or carcinogens in cigarettes. The difference with these substances is that there are direct, provable links to cancer. No vague, unrealistic numbers which prove nothing but how to practice bad science.

Fixing the Problem

Let us review some possible reasons why so many people believe such junk science and fall for deceptions and frauds. The teaching of science has taken a back seat at the elementary level. If classes get to science topics at all, it is only brief. Long gone is the day when science held its own with math and reading. High school is not much better with its often out-of-date books, which never get finished during the year. It is not just science, all subjects are shortchanged in many schools. A lot of these problems can be avoided by a very simple concept that is not unique to science: Test what you hear and do not automatically accept everything as fact without thinking it through. This is called common sense.

Will throwing more money at education solve this problem? We spend hundreds of billions on education, yet past generations of students in American schools used to lead the world in all fields with a fraction of the money spent today. It should seem obvious that the money spent today is not being used wisely. Students from private schools often score better because those schools have no choice but to spend every dime wisely. They do not have tax dollars being funneled to them regardless of performance with the possible option of raising taxes for more funds. Privatization of public schools is abhorred by unions, all at the expense of the students. Nor does it cost millions to set up standards that ensure children do not enter the world unable to determine fiction from reality. A sobering reminder of this problem is that many cults have lead their followers to certain death simply because their members had not been taught how to determine truth from deception.

Perhaps we should clarify that the science part of education is not a "cure-all" for all our problems. Science does have an important place in society and education, which is why it is at the focus of many of the discussions in this book. But be wary of those who make science sound like it has the power to accomplish everything or is the "savior" of mankind. This is scientific materialism which can turn science into a tool to convince (or coerce) you into believing something that may or may not be valid. Scientific materialism also abuses science by rationalizing to particular conclusions in order to support a person's beliefs or ideas. This is where science becomes the religion of scientism, which is nothing but a form of naturalism: Science is all that there is and can do it all! This is not science. It is the abuse of science.

The lesson from these problems is that people must recapture critical thinking skills. Every day new frauds, tabloid revelations and deceptions of all kinds are thrown at the public. Do not let your mind be abducted. A little educated thought and reasoning can be helpful to identify possible frauds. Be careful and look for fallacies or questionable reasoning. Ridding such nonsense from society is not inherently difficult and, in some cases, might save a few lives.

6

What About Those Aliens?

This is what it always comes down to: Aliens. They seem to come into the picture at some point. Always. Or at least most of the time. The Bermuda Triangle and Area 51 are crawling with aliens. Elvis may be one. They said so in *Men in Black*. People attribute to them Stonehenge, the Pyramids and all other ancient structures. Archaeologists and engineers are fairly certain aliens were not needed for such structures,[1-2] although it makes for a good movie as in *Stargate*. We have already taken a look at how easy the vast majority of "alien accounts" can be discarded as having nothing to do with extraterrestrials. Do *real* aliens or any form of extraterrestrial life actually exist in the universe?

When most people look into the night sky, they often wonder if any of those thousands of star systems harbor life-sustaining planets. With trillions of stars, there have to be millions of civilizations, right? You will probably be surprised to hear that this (abused) statistical argument does not have much support from science.

Not so Unimportant After All

The revelations from science in the past century, especially in the past thirty years have made some interesting discoveries. In order for life to exist, hundreds of interconnected constraints must be met in all the details of the universe. Slight variations of these constraints, in either direction, render life impossible.

Let us look at some examples. A planetary system, such as ours, located too close to a galaxy's center would receive too much radiation. Too far away and it would not have the correct quantities of materials (elements, etc.) needed to allow the planet to exist and life to survive. More than one star in a solar system and the gravitational forces and radiation become hostile to life. In a planetary system like our own, with one star, the star must have average size, age and luminosity (these variables influence such things as heat and radiation output). A variation of any of these properties either increases forces hostile to life or decreases the resources (elements, compounds, etc.) necessary for life. Already, we have eliminated billions of candidate solar systems for life.

The planet itself must be at a particular distance from the star. Its size must be just right for gravity's sake. Too much gravity and deadly gases (those with low molecular weight such as hydrogen) would not escape into space. If the planet is too small, the lack of gravity would allow oxygen and water vapor to escape. This was one cause in the loss of what could have been a life-sustaining atmosphere on Mars. Interaction with other planets and natural satellites also must be just right to avoid destructive tidal forces from interacting gravity fields (This is demonstrated on Jupiter's moon Io which is extremely volcanic due to being gravitationally "torn" between Jupiter and its outer large moons). Large planets such as Jupiter seem to protect inner planets from most "planet-killer" impact events caused by asteroids and comets by deflecting or absorbing them.

Even life-giving oxygen itself cannot vary much from the 21% content found in Earth's atmosphere. Increase the oxygen and its volatile properties become apparent. Fires would burn uncontrolled and oxidation would drastically shorten our life spans. Decrease it and combustion, the basis for technology and industry, would be difficult. And yes, large beings such as humans would have a difficult time breathing.

The nature of water is unique among substances in the universe. Its solid phase is less dense than its liquid phase. What if ice was heavier than liquid water? Bodies of water would freeze from the *bottom*, instead of just on top. A vital part of the ecosystem would die creating a cascade effect ending all life. Earth would have never made it through the Ice Ages, if it had made it that far. Other thermal and structural properties not only make water unique, but necessary to life. If this oddball of nature was only slightly different in many of its properties, life could not exist.

There are dozens more of these constraints that cannot vary significantly. Everything from the number of continents and their location to the frequency of earthquakes are factors. Too many earthquakes create obvious problems for life. Not enough earthquakes and nutrients required for life are all washed into the ocean and stay there. Earthquakes and continental plate movements recycle these nutrients, returning them to the surface. Also, life *must* be carbon-based. Physics, chemistry and biology seem to provide for no other option as the basis of life.

Science fiction often portrays life made from other elements or environments. Chemistry shows such depictions will remain in the realm of fiction. Other than carbon, only two elements exist that can support complex molecules, boron and silicon. Boron is extremely rare in the universe and is toxic at certain concentrations. Silicon-based molecules never reach the required complexity needed for information-rich life.

The same scientists who search for life in the universe have discovered these constraints. They had great optimism for finding many life forms in the universe and many still do hold out hope. Our intuition tells us there must be life throughout the enormous universe because of the vast number of stars. Abused statistics and wishful thinking ultimately must yield to the laws of physics.

What are the chances for carbon-based life then? One astronomer calculated the chance using 128 parameters[3] that must be met for life to exist and found that *one* planet with life in the *entire* universe would be extremely rare. In fact, rare is defined here as less than 1 chance in 10^{144}. Only a year later the known limiting parameters[4] increased to 158 and reduced chances to 1 in 10^{174}. As a result, one can safely say one such planet per acceptable galaxy is not only pushing it, but a bit outlandish. By calculating even a few parameters, we should not be here unless we are the *intended* result of these parameters.

What about those planets astronomers keep finding in other solar systems? None have been Earth-like planets. Most are in solar systems inhospitable to life. Astronomers keep finding more dead worlds. What about here in the Terran Solar System? Did Mars ever support life?

Liquid water may be rare beyond Earth, but it is not unheard of in the universe. Comets contain water. NASA's Lunar Prospector probe seemed to indicate that the Moon might have ice at its polar regions. What water is not, is an automatic indicator of life. Life needs water, but water is useless without the right environmental conditions.

Mars did have liquid water at one point. Recent studies from NASA probes orbiting Mars are now suggesting that this warm, wet period was geologically brief and environmentally turbulent (unfriendly to the development of life). Most of the channels, canyons and other water formations seem to have been formed by catastrophic flooding caused by impact events. The impacts released the ice-bound water, which did not survive long in Mars' weak gravity and thin atmosphere. The solar wind emanating from the Sun had already been stripping away Mars' atmosphere, including water vapor, which its low gravity was unable to prevent. What is left of the water is frozen as ice or found in clouds.

Mars' cataclysmic history seems to be bad news for finding remnants of life. What about the meteorites identified as originating on Mars? Out of about twenty meteorites found on Earth from Mars, one found in Antarctica *may* have fossils. But that is a hotly debated topic among scientists with most no longer supportive of the objects being fossils. Instead they are more likely to be inorganic formations. Even if we do find life, one may want to bet money it turns out to be from Earth. The planets have shared quantities of material through major impact events, which is how the famous Mars meteorite arrived here in the first place.

How and Why is Earth Here?

The conclusion to which science leads us is that the structure of the universe dictates carbon-based life and there is essentially nowhere for it to exist *except* on Earth. These facts obviously have not, and probably will not, change the minds of UFO chasers and sci-fi fans. It has depressed legitimate exobiologists, some of whom hope E.T. will save mankind from all its problems. However, do such sobering realities take away from the wonder of the universe? Should we stop exploring the universe? Does looking into the night sky mean earthlings should feel alone and depressed? No, no and no. If anything, some far-reaching questions should be simmering in your mind.

Why does the universe seem to be designed only for the carbon-based life we see on Earth? And how did so many interconnected constraints necessary to life arise in such precision? Mathematically, chance alone could never have produced such precision. It cannot even come close. Consider this: All of those parameters must be just right for Earth *to exist as it does*. The old paradigm that we are of no consequence in the universe and are just a random "pale blue dot" is no longer defensible. The universe must exist *as it does* for Earth to be here. To put it another way, the universe's structure seems to specifically serve as the foundation of Earth's existence. At the same time, the universe's structure seems to make the probability of other life existing impossible. *The bottom line is that the universe is structured for life to exist here on Earth and nowhere else.* This has some significant implications for the worldviews of naturalism and theism, which we will return to in a few chapters.

Profound thoughts for simply looking at the stars. But even looking at the stars can be confusing and full of myth and misinformation.

7

Figuring out Science

Soon after Sputnik was launched in 1957, the United States undertook an ambitious program to make its students the world leaders in the sciences. The minds behind this country's modern cutting-edge technologies were products of this education.

Nearly fifty years later, any discussion concerning education is likely to include talk about science textbooks riddled with errors. Or students who cannot compete with other nations and the lack of scientifically or technically educated persons available for high-tech jobs. All levels of schooling suffer from a lack of science instruction or science of any real rigor. Unless you major in a science-related field in college, or have an avid interest in science, you are out of luck. The media is full of bad, sound bite science. Even cable channels dedicated to education like the Discovery Channel often feature watered down science. Receiving a science education from the media is often only marginally better than using tabloids as your science texts.

"I suppose you want me to read a ton of books?" Well, are you not reading this? Alas, this old-fashioned method is still the most useful and detailed way to study science. Many scientists endeavor to bring science to the popular level through books. That does not mean that one should take everything verbatim and not test what you read according to the Rules and Maxims. There are plenty of bad books out there, but books are still your best bet at amassing knowledge about science or any other subject.

Generally, the sciences are not inherently difficult. Sure, highly specialized fields and their complicated mathematics may be difficult (or frightening), but that does not preclude the understanding of basic concepts. The goal of this chapter is to set you on the right track. Other chapters in this book have or will discuss other science-related topics, but this chapter will focus on some common myths and misconceptions that concern astronomy and space knowledge.

Earth's Moon

The Moon has always been a great source of wonder to mankind. In spite all our knowledge, and the fact that humans have visited the Moon, it is still a much misunderstood planet. For example, the Moon does rotate on its axis. Many people believe otherwise because we only see one hemisphere from Earth. We only see one side because its rotation is equal with its revolution period around Earth. If the Moon did not rotate we would be able to see all sides. This also means that there is not a perpetual "dark side" on the Moon.

> **Term Clarification:** *Rotation* is the movement of a planet on its axis that creates its day. *Revolution* is the movement of a planet around a parent object (such as the Moon around Earth, or Earth around the Sun), which is usually the basis for what is defined as a year for that planet.

Here is an experiment to better understand the motion of the Moon that is often suggested. Take a ball and slowly move it around a globe. Rotate the ball at the same speed on its axis as the speed it moves around the globe. Imagine people on the surface of the globe. Once your ball returns to its starting point, you will realize that the globe people have only seen one side of the ball moon as it moved around their world. Placing a lamp close to your imaginary planets will demonstrate lighting and phases of the ball moon. Now let us clear up some Moon terms that may be a source of confusion.

A "blue Moon" is not blue at all, but is when a full Moon occurs twice in one calendar month. On rare occasions, the Moon could look bluish when the atmosphere is full of smoke or ash. The term "honeymoon" may have come from the fact that June's Moon is often more yellowish from the heat and humidity in Earth's atmosphere—and June is the traditional month for many weddings. The "Harvest Moon" is the full Moon closet to the autumnal equinox (this full Moon usually occurs in September, but sometimes in October) that allowed farmers extra light for harvesting. The "Hunter's Moon" is the full Moon of October allowing extra hunting time back in the day. Granted, these terms applied to a time before tractors with headlights and hunting regulations, but they are the best remembered of dozens of Moon names.

How about the old adage that a ring (or halo) around the Moon is an indicator of coming bad weather? The ring is caused by ice crystals (in thin, higher clouds) or water droplets (in lower clouds). This is not always an indication of coming precipitation. However, if the phenomenon persists through the night, then it is a good indicator.

The words lunatic and lunacy take their root from the Latin word for the Moon, Luna. Does the full Moon cause an increase in crime and craziness? Do more births happen during the full Moon? These questions should be easy to answer by simple statistical analysis (such as counting crimes committed or births that occurred). What such studies show is that there is not an increase in crime or births. So why do many people believe otherwise?

This goes back to the selective evidence and confusing correlation with causation fallacies. If you were to record births at only one or two hospitals, maybe you would get lucky. Record births at a much larger cross-section of hospitals, such as *all* of them, and you will find your theory out the door. Small statistical samples can be used to prove just about anything. People like to find a small sample that supports their belief and use it. However, the larger the sample, the more indicative of reality. That is why polls of small groups of people should not be referred to as a source of authority without carefully controlled conditions.

Why are numerous phenomena attributed to the full Moon and not other Moon phases? After all, the Moon's *actual physical size* is *not* changing. Regardless of the phase, the same mass is always there in orbit. The self-fulfilling thinking that often plagues the brain also comes into play. How often do you notice the Moon when it is *not* full? You are more apt to take notice of a full Moon and any events that *happen* to occur that night. People also tend to remember the hits rather than the misses. A night that "proves" your theory tends to stand out more than one that does not.

"What about the Moon's gravitational influence? It creates tides and pulls on the surface of Earth, so couldn't this influence people?" Not exactly. Gravity works over large surfaces rather than on specific points. Consider that the Moon exerts its force over the surface of the Earth, which is about 8000 miles across. People are only about one foot thick, give or take, so there is significant less surface area for the Moon's gravity to effect. As far as the Moon is concerned, you are but a speck on Earth's surface (We will discuss gravity in more detail shortly in our discussion of general relativity).

This also means that the close coincidence between the female menstrual cycle and the Moon's synodic period (the time between two full or two new Moons) is unrelated. Additionally, only one other mammal has a similar cycle and not all women are on the same cycle. Nevertheless, menstrual comes from the Greek word for the Moon, menses, and many mythologies refer to the Moon as a female—such as the Roman Luna and the Greek Selene Moon goddesses.

Sky Watching

Why is the sky blue? No, it is not the reflection from oceans. If this is true, why are the inland skies of Nebraska blue? As the white light from the Sun enters the atmosphere, it is broken up into its constituent colors as it bounces off particles. The colors with shorter wavelengths, such as blue, scatter more in the atmosphere. At sunset, the light has to travel through more atmosphere (because the Sun is at a lower angle relative to your observing position), thus the blue light is absorbed by more atmosphere, leaving more reds and oranges, which have longer wavelengths.

Your astronomy knowledge may include a couple other misunderstandings as well. Earth is further away from the Sun in the summer than winter, not vice versa. It is Earth's axial tilt that creates seasons. As different hemispheres receive the direct rays of the Sun throughout the year, the seasons change.

Polaris, the North Star, is not the brightest star in the night sky. It barely makes the top 50, but it will lead you in the right direction. "Light-years" is not a measurement of time as in "we are light-years ahead of the competition," but is a distance measurement. One light-year equals 6 trillion miles, the distance light travels in a year.

Theories of Relativity

Some people blame Albert Einstein for relativism. No, he formulated the theories of special and general relativity. These have to do with the physical nature of the universe, not the philosophy of relativism. He preferred the less confusing and more accurate term "theory of invariance" because the unchanging speed of light is so central to his theories. The naming of things in science, however, is not always a scientific process. First, let us examine special relativity which concerns itself with frames of reference, $e=mc^2$ and the constancy of the speed of light.

Consider the principle of relativity. If you are standing in the back of a pick-up truck going 40 mph and throw a ball forward at 10 mph, the ball is only going 10 mph from your perspective. Someone watching the truck pass by will record the ball going 50 mph, the speed of the truck and the ball added together. Both of these people record different speeds based on their position, but both are correct. Keeping in mind their frame of reference resolves the paradox. This is essential to one of the components of special relativity: The laws of physics are the same to everyone, but you have to keep in mind relative motion and position.

This is not particularly hard to understand, but it is about to get a little unintuitive.

To an outside observer, the speed of the light emitting from the headlights of the truck is *not* equivalent to the speed of light added to the truck speed. If the truck driver tried to chase after a beam of light, no matter how fast he went, the light would still retreat at light speed (186,000 miles per second). Light travels at this speed from the moment it is emitted from a source: No acceleration period to get to that speed. These are measurable, observable results. The speed of light does not change under natural conditions, it is constant in any frame of reference (light can slow down through a substance, just like sound, but the speed of the light at its creation is always 186K mps).

In Einstein's mass-energy equivalence equation, $e=mc^2$, "c" is the speed of light. Mass and energy can be converted into each other, which is fundamental to fusion in stars and the very structure of energy and matter (and spectacularly demonstrated in a nuclear weapon, which converts a relatively small mass into a very large energy output). This also makes it apparent that the speed of light cannot change or else the fundamental structure of our existence, which this equation describes, ceases to exist. The speed of light is another one of the fine-tuned constants discussed in Chapter 6 which determines whether or not we exist. Also, some simple algebraic manipulation of this equation shows why no object with mass can exceed light speed.

General relativity is largely concerned with the dynamic nature of the universe (more on this Chapter 10) and the nature of gravity. Gravity is one of the four fundamental forces. The other three are the electromagnetic, strong nuclear and weak nuclear forces. The latter two deal with atomic structure (holding everything together), rates of radioactive decay and so forth. Electromagnetic force is what holds atoms and electrons together. Gravity is the weakest of the four. Think of it this way: A magnet will pick up a heavy piece of metal *against* the force of Earth's gravity.

Gravity is not projected from a mass, but is a result of the mass warping the space around it. This can be difficult to understand, because space is hard to imagine as anything other than, well, space. However, space, or the fabric of space, is part of the universe's structure. We can actually observe this fascinating characteristic of space through astronomical observations of a phenomenon known as gravitational lensing. This occurs when we see two images of the same distant object. The distant object is situated directly behind a closer, more massive body such as a galaxy (note that the distant object is not close to the foreground object). The mass of the galaxy bends (or warps) the space around it, allowing the light to travel around the galaxy and form a second image of the source object in a different location than the primary image.

Both relativity concepts describe a unique property of time: It can slow down! The time on an object traveling at very high speed appears to run slower as measured from an observer's clock (the observer is not traveling at the high speed). The person on the fast object does not notice any rate change on his clock, but when he returns from the high-speed trip, he would find that the world had aged more than he did. At high speeds approaching the speed of light this difference in the passage of time may be recorded as decades or centuries. In the life of the average human, this effect is imperceptible. For an astronaut traveling around Earth at high speed (17,000 miles per hour or so), the difference is measured in fractions of a second. This effect, known as time dilation, is predicted by special relativity.

In terms of general relativity, time is slowed down by gravitational fields. If you could hypothetically travel through the immensely strong gravitational field of a black hole without dying, when you return to Earth you would find that more time has passed at home than in your spaceship. Note that the slowing of time you experienced in both examples is *actual physical time*, not your perception of the passing of time. In other words, time seemed to pass normally to *you* in your context, but time was actually physically progressing at a different rate than it was for people in another context. Comprehending these concepts is in part contingent on understanding time as a dimension—like height, width and length—which is addressed in Appendix A. Together these four dimensions are often referred to as *spacetime*.

There you have some of the basics of both relativity theories in a nutshell. As described, these "theories" are testable. They may be the most tested concepts in physics—and they supercede Newton's Laws of Motion and Gravitation which break down at a certain point—but many people (although less and less do) still refer to them as theories. On the other hand, even something referred to as a law is incomplete. We never can break down something completely. There is always room for refinement and more precise tinkering. This is the essence of what is known as Gödel's Incompleteness Theorem. However, when a law or theory is confirmed to such high precision (to at least fourteen decimal places in general relativity's case[1]), it is a law or fact for all intents and purposes. Since this all may be giving you a headache, we will look at only one more example of confused science.

Ozone

The controversy over the ozone hole is perhaps the best example of bad media science. As soon as the ozone hole was discovered the media began predicting the doom and gloom that would follow. Increased ultraviolet (UV) radiation can increase skin cancer cases and damage plants, but is the hole caused by mankind?

Ozone, which protects us from most of the UV radiation emitted from the Sun, is the result of chemical processes caused by the UV radiation itself. When the radiation decreases or increases, the ozone does the same (see why below). These variations are caused by cyclic weather on the Sun. This weather causes fluctuations in ozone far greater than is possible by manmade chemicals. This does not mean we should start pouring CFCs back into the atmosphere. It does mean we should not be claiming Armageddon when basic chemistry suggests restraint.

> **Brief Chemistry Lesson:** What is ozone? Ozone is a form of oxygen. The oxygen we breathe is O_2, ozone is O_3. When O_2 is broken apart by UV rays, it recombines with another oxygen molecule to form O_3. Hence, the more UV rays entering the atmosphere, the more ozone that is produced. Electricity can be the catalyst for this process as well. If you smell something a bit sharp when running your electric train, it is probably the sparks creating ozone. Ozone in large quantities is toxic and is a major component of smog. In the upper atmosphere, it absorbs dangerous UV rays.

Why Bother with Science Anyway?

The importance of science should be evident. It allows us to understand the universe around us and distinguish between reality and fantasy (although that does not mean it has to prevent us from enjoying science fiction). More importantly, scientific knowledge instills critical thinking and reasoning processes. This reason itself should be enough to put science back at the forefront in the educational system. However, for many people, the wonder of understanding is reason enough.

Even the importance of science does not prevent it from having a contradictory position in our lives. In spite of its usage in technology, medicine and elsewhere, many are suspicious of science. "How can people know these things? How can they know how far stars are? We can't know what happened in the past!" These attitudes often result from a simple lack of scientific knowledge further fed by sound-bite science on the news which explains little and assumes a lot. When one actually studies the sciences, or finds a resource that explains a particular subject well, the realization that science is not mysterious or mystical is often a pleasant surprise. Sometimes it is just a matter of remembering what you learned in that high school science class you took so long ago.

Again, we cannot turn science into scientific materialism, making it out to be all-powerful and all-knowing. Then science becomes god-like and allows mistakes or personal philosophies and beliefs to masquerade as facts. The understanding of science can improve our lives, save others and protect the world for the future. That is if we can differentiate between real science and radical philosophy pretending to be science.

8

Environmental Protection or Ecoterrorism?

At the beginning of the Twentieth Century, nature enthusiasm began to sweep the nation. Citizens were tired of city life and trips into the wilderness became all the rage. Out of this fad would appear a new appreciation for the wild environs of America. This in turn led to a new national park system and various wildlife protection laws. America wanted to preserve its lands for future generations.

Fast-forward a few decades to the 1960s and 1970s to a new environmental movement. The movement produced some notable results such as cleaner air and purer waterways. But it also produced a number of illogical ideologies that often cloud environmental issues. Perhaps it was the other way around, and these ideologies stemmed from, or fed off of, our old friends naturalism and relativism.

For some people, it is no longer enough to plant trees and preserve forests. Some groups want the public to abandon the entire lumber and paper industries and all of industrial society. Humans are parasites and they want nature returned to how the first Europeans found it. One problem with this reasoning is that early pioneers did *not* find untouched lands. They were surprised to often find young forests. It seems many natives loved to clear forests, especially by burning them down.

What the extremists tend to forget is that many forests are considered an agricultural crop. Grown and cared for specific purposes in the same way as corn or wheat. Many forests are protected by the government for the sole reason of saving them for future harvest. Nature is protected, not destroyed, and human lives are improved.

Some extreme environmentalists will do anything to stop the harvesting of trees, even at the risk of human lives. This results from transposing human emotion to the death of plant life. They equate the importance of trees with that of human life or sometimes assign greater meaning to plant life. Biology, on the other hand, confirms plants are not sentient beings like humans nor do they approach animals on the emotional level. Our senses confirm that trees and their relatives do not approach, by any stretch of the imagination, the complexity and consciousness of humans. It seems that for some people, emotions and personal beliefs are replacing physical reality. Some radical environmentalists have gone as far as "spiking" trees, which can injure, maim or kill loggers when they cut into the trees with their chainsaws. Yet, when all is said and done, these radicals still eat plants to survive.

If some people elevate plants to such a level, one may correctly assume that the same is sometimes done with animals. Consider the term "animal rights." How can creatures without anything remotely close to higher consciousness have rights? People also try to equate the abilities of some animals with advanced intelligence. At best, most of these animals learn by copying what they see or from some degree of instincts or limited intelligence. Never do we see intelligence or emotion of animals begin to approach that of humans. Even the vaunted dolphin has been found at times to be a violent, uncaring animal in the wild. This does not mean that animals should not be held in higher regard than nonliving matter. The point is that there is a significant difference between the reasonable treatment and uses of animals as opposed to elevating their importance over human life.

The radical animal rights group, People for the Ethical Treatment of Animals, or PETA, ran a campaign not long ago that said our young people were better off drinking alcohol than milk. Yes, alcohol is the substance that kills many people when it is abused and has very little nutritional value (what value alcohol does have is greatly offset by the side effects of abusing it—with many people in this country seemingly unable to consume it responsibly). Milk, its well-known nutrition notwithstanding, comes from cattle. Hence PETA claims the cattle are being misused and abused to obtain milk. Farmers will tell you that abused cattle do not produce much milk. Even members of PETA were upset over the "alcohol campaign" which, combined with the disgust of the public, caused PETA to pull it (yet PETA has since tried to revive it).

Emotional pleas have also led to attacks on the food industry and hunting. Such pleas are often inconsistent coming from animal activists who use animal products or extreme "tree-huggers" that nevertheless eat plants. We are also told of the "horribleness" of hunting, but are not given any alternatives to population control. Moving animals only moves the problem. Arriving in a completely new area does not do wonders for their survival either. Few realize that until hunting regulations were established, many animals were on the verge of extinction. Now states like Pennsylvania have more deer than they can support. Do we let them starve and wreck cars or do we responsibly harvest them? Some claim hunting is cruel. Such claims are made by people who obviously do not understand starvation or being hit by a car. Deer no longer have natural predators in most regions of North America, so humans take their place. It is either hunting or the reintroduction of mountain lions.

What needs clarification is that there is a difference between responsible environmental management and what amounts to "nature worship." One is free to worship as they please. However, developing such beliefs that place the importance of nature over humans is dangerous. When humans are no better than animals, society slips into relativism. Human life then loses importance and anything goes. Some of the results were discussed in Chapter 4.

If keeping animals as pets, raising others for food and doing so responsibly, is as evil as these groups claim (actually many of them keep pets, yet hate zoos), how can we survive? What can we eat? What of the animals that kill and eat each other? The thinking of the radical environmentalists is very inconsistent. If some of these people are willing to endanger human life over flora and fauna, what kind of message does that send?

Extreme Naturalism

It sends a message that humans are no more important than anything else on this planet. So why do we bother punishing murder, terrorism or genocidal tyrants? This form of environmentalism is naturalism taken to an extreme: We are just another animal evolved from some sort of primordial soup.

Maybe you are thinking I am making too much of the some of these extremist groups. If so, consider the following quotes made after the September 11th terrorist attacks:[1]

> For 35 million chickens in the United States alone, every single night is a terrorist attack.—Karen Davis, United Poultry Concern

> Worldwide, every day, 125 million innocent, sentient animals are dreadfully abused and butchered for food. These tragedies are perpetrated by a worldwide animal agricultural terrorist network that is much more threatening to planetary survival that the Al Queda network.—Alex Hershaft, National Chairman, Animal Rights 2001

> …Animals need saving and that's more important…This New York thing is being blown out of proportion.—Lee Ryan of the British pop band "Blue"

These illogical philosophies have brought much suffering into the world. When will their supporters—self-proclaimed scholars and intellectuals—stop trying to promote them? Sooner or later, most will have to face the glaring inconsistencies of their thinking.

9

Testing All Things

Science can purify religion from error and superstition; religion can purify science from idolatry and false absolutes.—Pope John Paul II

We have taken the Maxims and Rules of critical thought and applied them generally and specifically to a wide variety of topics. Now it is time to focus in on naturalism and theism. The following chapters will explore their relation to, and influence of, science. We will first turn to the dominant form of theism, Christianity, and its long storied relationship with science.

"Science is a matter of observation, data and fact. Religion is a matter of faith and belief.[1]" This statement sums up what many people believe. It is only partially correct. The part about science is correct, but the part about religion is a broad stereotype. One could argue that many religions could be accurately described by that stereotype, including some under the heading of theism. Many would argue that orthodox Christian theism (see Condensed History Lesson below) does not fall into this stereotype and is, by definition, based on reason.

"Wait a minute! I thought religion is about having faith. Faith is intangible. People tell us you just need faith to believe. It isn't about reason and reality." That is how many people approach faith. It is not, however, how Christianity has ever defined faith.

Blind Faith or Reason-infused Faith?

In fact, Christian doctrine derived from the Bible defines faith as having two components. The emotional (from the heart or soul) and the intellectual (from the mind). A person takes a "leap of faith" with their heart first—and lives in faith by trusting God—but solidifies and defends this with their mind which is attested to in such verses as Matthew 22:37–38 and Romans 12:2. This is taken further by Peter in 1 Peter 3:15 who implores followers to give a reason for what they believe. In 1 Thessalonians 5:21 we read a command to test everything and such testing is encouraged further in Job 34:4, Acts 17:11 and Revelation 2:2. Since Christianity is based on the Bible, we can assuredly assume that these verses can be used (and they are) to define Christian faith.

Think about what these fundamental Christian doctrines are saying. They are telling us to practice the same critical thought that we have been discussing. The command to test everything and give a reason could not be clearer. It is also unique to Christianity. No other religion so boldly asks its people to test its beliefs because this could be applied to the religion itself. The way the Bible defines faith (biblical faith) is different from what most people practice (blind faith). One could live all your life by this latter version of faith, but that makes it harder to fulfill those beliefs and biblical mandates. It also opens the door to aberrant theologies and illogical beliefs. Christianity has never demanded that people separate faith from reason. It states that each is closely bound to the other.

The Bible lays out further guidelines for critical thought when it warns readers about being taken in by deceitful teachings (1 John 4:1), human traditions (Colossians 2:8, Isaiah 29:13), emotions (Proverbs 28:26) and believing in myths (1 Timothy 1:4, 2 Timothy 4:3–4, Titus 1:14). Most of the Book of Proverbs implores readers to search out wisdom. The modern scientific method, another core of critical thought, can trace its roots in biblical study.[2] Apparently the Bible insists there is an absolute truth standard just like the one we derived from logic. What sets the Bible apart is that it claims it comes from the source of that truth and then it invites you to test that claim.

It is beyond the scope of this book to address most of the claims of skeptics and critics against Christianity. The point of this chapter is to determine if Christianity is really antireason and antiscience. Many scholars have committed volumes of works over the centuries to adequately address other debated matters. This is the field of apologetics. Apologetics is derived from the Greek word apologia, which means to defend something. Apologetics is what Peter was talking about in the before-mentioned verse, 1 Peter 3:15. Some resources on apologetics are listed in the Notes for those who want to pursue the subject further.

If you are thinking many do not follow these biblical mandates for reason and testing everything, you are right. Christianity's abandonment of this intellectual pillar is an epidemic problem that largely took root in the Twentieth Century.

Nowhere is this more obvious than in the "religion is antiscience" perception common in society.

Christianity, the Antiscience?

It was not always this way. History is full of great minds that were Christians who contributed to the worlds of science and philosophical thought. Newton, Kepler, Copernicus and Galileo revolutionized science because of their motivation to learn about God by studying creation. Georges Lemaître (first proposed the big bang theory), Angelo Secchi (father of astrophysics), Gregor Mendel (father of genetics), Roger Bacon (father of chemistry) and Albert the Great (father of geology) were all priests or monks.[3] The list of Christian persons who are part of our scientific and intellectual heritage is much longer than this. It was once not uncommon for those involved in ministry or members of the clergy to be well schooled in the sciences. Why then do so many claim science and religion are separate realms?

Part of it comes from a fear that science will, or does, prove religion false. People who think science does "disprove" belief in God usually promote the "science and religion are incompatible" mindset. They often quote the same one or two "incidents" in Christianity's history as "proof" of their belief. The first event they point to the Roman Catholic "persecution" of Galileo. Galileo published his ideas (supported by observations) that Earth was not the center of the universe. This was not a new idea: Copernicus is usually credited with it, but Aristarchus had proposed it centuries before.

The Roman Catholic Church had misinterpreted the Bible in thinking it claimed Earth was at the center of the universe (also note that Protestant reformer Martin Luther was highly critical of Copernicus' heliocentrism before the "Galileo Affair" took place) and ruled twice against Galileo. It was not until after the second time, when Galileo was 70 years old, that he was placed under house arrest. This erroneous geocentrism was supported by the works of Plato and Aristotle, which were held in higher regard than their largely forgotten, but correct Greek brethren, Aristarchus. Galileo was probably never in danger of losing his life by the Inquisition, as many believe, and his "recanting" of heliocentrism was but a formality to get Church officials off his case while he continued his work.

Another incident pointed to is that of Giordano Bruno who was burned at the stake. This punishment is often partly attributed to his scientific beliefs such as there may be life on other worlds and that the universe was infinite. Critics hold such examples, as rare as they are, and forget the extensive history of Christian leadership in the sciences. Even in the same chapter discussing "antiscience religion" some will discuss the achievements of Newton and others. Perhaps scientists should brush up on their history.

The Roman Catholic Church continues to support scientific studies to this day (among other things, it maintains observatories), but without the conflicts of the past. Over all, however, most of Christianity does not study science at any significant level. Seminaries across the board focus little on science. Those that do are often infected by naturalistic philosophy. Some Christian supported colleges and parachurch organizations are nearly the last vestige of scientific study in Christianity. The long tradition of accepting general revelation (nature or science) as equivalent to special revelation (the Bible) and studying the former to enhance the latter became a causality of the "religion is antiscience" paradigm. The Bible speaks to the validity of general revelation in many verses including Psalms 19:1–4, 50:6, 85:11, 97:6 and 104. Both types of revelation have different roles, but both have equal validity as *truth*.

Some say, "Science cannot be on the same level as the Bible!" I answer that by asking, "If God inspired the Bible and created the universe, shouldn't both agree? How is one *truth* from *God* more valid than another?" That is the logical problem into which some creationists get themselves into, as we will see in later chapters. They will promote antiscience positions against well-proven science while at the same time presenting their own science as a proof for their theories. This "science is sometimes good, sometimes bad" thinking has to be confusing to many. This is completely unnecessary if one studies science using the methods of reason discussed in this book and by the following logical foundation: "If God created the universe and the Bible, then both should agree. Any disagreements are due to misunderstandings in our interpretation." That is a piece of logic often tossed out the window.

It is obvious that critical thought was thrown out the window in the cases of Galileo and Giordano. For the record, the Bible does not support geocentrism. The verses that were said to support geocentrism are Psalms 93:1, 104:5 and Ecclesiastes 1:4–5. Galileo rightly pointed out that these verses need to be interpreted according to the frame of reference in which they were written.[4] That is how all verses should be read and it is also included in the first step in the scientific method. To clarify, this means that one should determine what was the frame of reference or the context (i.e. historical, geographical, etc. or use of literary styles such as symbolism, etc.) of the writer.

A closely related misinterpretation is suggested by people who claim the Bible promotes the idea of a "flat Earth." The belief in a "flat Earth" was not as widely believed throughout history as you may have read, nor did Christianity promote it. This myth has largely been propagated through poor biblical study and by some unscholarly "historians" and skeptics. These skeptics base this claim on the "four corners" or "four quarters" in verses such as Ezekiel 7:2, Isaiah 11:12 and Revelation 7:1. Most thinking people, however, understand this to be symbolism or in reference to the cardinal directions (north, south, east and west). This becomes more obvious in Isaiah 40:22's usage of "circle of the earth." Job 26:7 infers this in stating Earth is suspended over nothing. A more obvious example is found in the Job 26:10, in which a more literal translation uses "circle" instead of "horizon." By doing so, Job predates the Greek "discovery" of a spherical Earth by many centuries.

Careful reading, as opposed to superficial reading, could solve or avoid many problems. A complete study on the Bible and what it may or may not agree with in the sciences is beyond the scope of this book. However, the coming chapters and appendices will add some more information to this topic.

Testing All Things

The point is that Christianity is not the evil, archenemy of reason and science. They have shared an intertwined existence for centuries. The "incompatibility" of the two is a construct of modern skeptics and people (including Christians) who do not test what they hear or are taught. This "science disproves God and/or Christianity" mindset is one of the principal challenges to modern Christianity, second only to defending the historicity of Christ and his resurrection (see Appendix F). The abandonment of its intellectual heritage (which includes more than scientific studies) by Christianity has led to more aberrant theologies, cults and denominational splintering in the past century than at any other time.

> **Condensed History Lesson:** Originally, there was one Christian church, or denomination. This one church seemed to have been the intention of church founders. The Church saw its first fracturing as one part centered in the western Roman Empire (in Rome) and the other in the eastern Roman Empire (in Constantinople). Their distance and other factors would eventually produce two churches, which would become known as Roman Catholic and Eastern Orthodox. Later, reformers would break away from the Roman church when the Church did not take well to theological and anticorruption reforms. Many early reformers did not want to split the Church, but that obviously happened anyway. So now a third major denomination, Protestantism was produced. Each of these "major three" can be divided into hundreds of other denominations. Orthodox Christianity is defined as the denominations that share all of

the same fundamental beliefs.[5] One source classifies these under the headings of evangelical Protestantism, conservative Roman Catholicism and the Eastern Orthodox church.[6]

These many denominations are often the result of cultural differences in ways to worship and disagreements on "secondary issues" rather than disagreements on fundamental beliefs. One troubling aspect is that many people look down upon other denominations thinking theirs is the best. They wear their denominational label like a badge of honor. They seem to miss the point that their "church-going" is supposed to be about being Christian, not Catholic, Methodist, Baptist, Lutheran or whatever they may be. This problem and the existence of so many denominations are a sign of the inability (or unwillingness) to work through, or critically analyze, problems. Instead of declaring one's traditions or beliefs right or wrong, why not determine which are allowable and consistent with Christian orthodoxy? There is room for some differences, but many of those differences should not be seen as dividing lines.

Perhaps we should look at the lack of "testing all things" at a very basic level. Over time people have believed many things that they "remember" the Bible stating, but in actuality they are remembering what *other people* have told them the Bible says. For example, we all remember there being three wise men in the Christmas story. However, the Bible does not say how many there were. We *assume* there were three because Matthew 2:11 mentions three gifts were given to the Christ child. It does not necessarily follow from this verse that there were three: At least two, but maybe four. Some traditions list many more wise men. We have *assumed* three for so long, we have come to believe the Bible says so, albeit wrongly.

Many times you probably have heard people repeat that the Bible says the lion will lie with the lamb, but it is actually the wolf with the lamb (Isaiah 11:6, 65:25). Delilah did not give Samson his famous haircut (Judges 16:19). No where does Exodus claim that Pharaoh drown in the Red Sea crossing (Exodus 14:23–15:21). This clarification undermines the claim of skeptics that Egyptian histories do not list any pharaoh dying during this time frame. Nor does the Bible record Jesus stumbling and falling when carrying the cross as some traditions claim. And the Bible does not say that Eve's "apple" was an apple at all (Genesis 3), but generically calls it a fruit.

Also, how many times have you heard the myth that women have one less rib then men? Both have the same number of ribs and critics use this to point out that creating Eve from one of Adam's ribs does not make sense. However, as

modern Bible translations often footnote, the more accurate rendering is probably *part* or *side* instead of rib. That makes sense in light of modern genetics,
which affirms men and woman share common origin.

These erroneous beliefs are fairly minor, yet they are often taught in Sunday
schools and from the pulpit. However, they establish a disturbing pattern. Entire
religions and belief systems have been created due to the lack of critical thought.
Mormonism was founded by Joseph Smith who claims to have received revelations from God in the 1800s. Mormons, or the Church of Jesus Christ of Latter
Day Saints, insist they are part of the Christian orthodoxy (mainstream Christians churches that share the same foundational doctrines). Those churches disagree because the "revelations" of Joseph Smith, recorded in the Book of
Mormon, are highly suspect. This book claims to reveal the history of ancient
civilizations in America that had Hebrew and Christian beliefs.

No ancient copies of this book exist. No persons, places or nations it lists have
been found. There is absolutely zero historical or archaeological evidence to support Smith's writings and there have been 3913 changes in the book since its first
printing.[7] Smith, who claimed to be a prophet of God, made 64 specific prophecies. Only six were correct.[8] Some of his prophecies included Jesus would return
to Earth by 1890 and the Moon would be found to be inhabited by six-foot tall
people.[9] Should not a true "prophet of God" be 100% correct? In fact, Deuteronomy 18:21–22 provides that very test: If what the prophet says does not come to
pass, he is not a prophet. Sounds like common sense.

As with many figures in history that tried to create new religions or philosophies, Smith was probably trying to justify (or hide) his own life, actions or
beliefs. His teachings on polygamy probably were created to explain or justify his
endless affairs with women (he ended up having many wives). Smith's early
teachings on marrying one wife conveniently disappeared. The Mormon church
no longer officially supports polygamy. Nevertheless, this gives significant insight
into the motivation and integrity of Joseph Smith.

Aberrant Science for the Believer

The Internet is known for its production of frauds and urban legends. Rarely a
week goes by without receiving at least one e-mail claiming something fraudulent. One common urban legend that often makes the rounds is that NASA has
discovered Joshua's missing day. This legend is nothing new, it has been around
since the 1970s. E-mail has given it new life.

Many are excited when they receive this biblical "proof." The only problem is that it is a fraud.[10] Every examination into the claims of this legend has produced the same conclusion: Someone made it up. Christian scholars hate frauds like this for obvious reasons. They argue there is plenty of *real* science that can be pointed to as evidences supporting Christianity. The general public, Christians included, however, can be quite gullible.

Another popular misguided biblical science "proof" is that of finding the gospel in the stars. Proponents of this theory claim the gospel is written in the constellations and was given to ancient man before written word, perhaps in Adam's time. Writers match up the constellations to the gospel accounts and claim evidence. There are some problems with this reasoning. First, all modern constellations were not always recognized as they are now. Every culture had different constellations it imagined or created. Some had similar constellations with each other, but not enough to prove a shared lineage to a gospel in the stars.

Secondly, and most importantly, the apparent position of the stars changes over the millennia. The axis of Earth changes its direction (pointing into space) as it completes one movement like a toy top every 26,000 years. This is called precession and causes the apparent positions of stars to change over time. Polaris, the North Star, has not always been in the northern sky, nor will it always be there. This and other galactic and stellar movements confirm that modern constellations did not even exist at man's origin.

One of the best-known myths concerns reports that claim Noah's Ark was found. The truth is that no proof of an ark has been found and some claims have been shown to be entirely fraudulent.[11] A dated piece wood supposedly from the ark has such wild error bars in the dating that it is worthless as evidence. Chapter 11 will discuss why looking for the ark on Mt. Ararat is futile, but for now consider some common sense. In a flood-ravaged area, would the people leave the ark to rot or use it to build homes?

There are other urban legends concerning "modern Jonahs" found in whales,[12] "evidences" of living dinosaurs (see Appendix D for more on dinosaurs), hell "discovered" in Siberia,[13] the "deathbed conversion" of Charles Darwin, "codes" hidden in the Bible[14] and "revelations" of all kinds from self-proclaimed prophets. These claims can be thrown into the trash bin marked "Frauds, Misidentifications and Delusions."

Restoring the Mind

Do not get me wrong, I am not trying to minimize the other side of faith. That side—the trust, the prayer, and so forth—is an essential part of Christianity. The point here is that the part that includes using your mind is often ignored. We are told not to rely on man's fallible wisdom (1 Corinthians 2:5), so this intellectual side of faith is essentially the search for ultimate, underlying, objective truth. We have discussed before how such truth must exist and that this existence requires a higher source. The two sides of faith are intertwined to the point where it is impossible to divorce one from another.

The trend of anti-intellectualism may be reversing in some cases as Christianity recognizes the problem exists, but it still remains a considerable problem. Too often people blindly believe or trust what they hear because it comes to them wrapped in a Christian context (from Christians, in a church or in a "Christian" book). This will remain a significant problem until the churches and their educational outlets (schools, seminaries and colleges) return to teaching critical thought and science. Christianity must return to the traditions of education and intellect begun in the Middle Age monasteries that helped to found Western Civilization and make the quest for wisdom foundational once more.

It has been said, "Peace cannot be purchased at the expense of truth.[15]" This is something that many should take to heart concerning all issues. In Christianity, this once meant not only passively defending what they believe, but being on the forefront of debating secular thought. Apologetics has *both* a defensive and an offensive component. Too often the fear of offending someone prevents resolution of problems. People who are truly interested in finding the truth will not be offended. Those who take offense usually are the ones who cannot defend what they believe. Christianity should not be afraid of defending its core of orthodox beliefs against errant denominations, beliefs and secular critics. Many secondary issues have room for interpretation within the pale of orthodoxy, but some (like the creation-date debate to be discussed shortly) need to be addressed to remove the criticisms of opponents.

Of course, these are the same problems (lack of critical thought, etc.) that the rest of society is having. One has to ask what would have happened if Christianity had not stepped back from debating secular thought (relativistic, irrational thought) and not abandoned its scientific and intellectual heritage? Would the societal problem of anti-intellectualism, and Christianity's own problems, have progressed as far as they have?

10

Getting a Bang out of the Universe

After finishing our discussions on bad science and the role of reason and science in Christianity, we will now focus on two case studies in this and the next chapter. These chapters concern the misunderstanding and misuse of science to support personal religious or philosophical beliefs and superficial biblical interpretation. They will also set the stage for the topic of origins that we will address next.

The big bang theory, or more precisely described as the big bang model, is one of the most often misunderstood theories of science, especially by certain creationists and skeptics. This is the result of not studying the model itself, replacing the science with personal beliefs and not testing what others claim.

First, a definition. The big bang "creation event" did not sound a "bang," it was a sudden *expansion* of space that *carried* matter and energy with it, beginning from what is called a singularity. A singularity has an infinitesimally small volume, whose "shape" may have been similar to a dimensionless point or plate (which "shape" is still not fully understood). A black hole is a type of point singularity we know to exist. No space, time, matter or energy existed before the big bang event.

Some young-earth creationists (who believe the universe is 10,000 years old or less) try to discredit the big bang or claim it is invalid. They do so because the big bang supports an ancient universe, which contradicts their interpretation of Genesis. Because of the big bang's support of an ancient universe, these creationists claim this model supports naturalistic biological evolution. On the contrary, as we will shortly see, neither the big bang nor ancient universe supports the theories resulting from naturalism.

Young-earth creationists often claim that the big bang violates the Second Law of Thermodynamics which states all systems have the tendency to increase in "disorder" over time. They ask, "How can the universe, with the order we see today, start out of this disordered 'bang?'" First we need to look at exactly what the Second Law states before we answer this question. The "disorder" is concerned with the dispersion of usable energy and this change in the level of usable energy in a system is known as entropy. The more energy lost (or "used") in a system (such as in a car engine), the more entropy increases, hence the use of the term "disorder." This disorder may not be so obvious in your car engine (unless you recognize that the heat and noise output are losses of the energy derived from the fuel), but throw a pumpkin out the window and its subsequent disorder becomes visually apparent even if you do not realize that it expends energy on impact.

On the scale of the universe as a whole, entropy measures how close the universe is to the point of equilibrium (no more possible energy change). And that is technically what entropy is all about. Cars and pumpkins are not isolated because you can always add more energy to a car or glue the pumpkin back together. Energy cannot be added or restored to the universe once it reaches an "expended" state.

Entropy can be too broadly applied under the guise of "disorder" to systems or processes that do not have much to do with energy dispersion. For example, while chance cannot produce highly complex organisms, this is not due to entropy. This is due to chance being a forceless entity. Complexity requires direct design or underlying constants that were ultimately designed. So the creationist question should be, "Was the big bang a disordered explosion?" It certainly was not.

There existed at the big bang event a state of order that was far greater in magnitude than the order we find today underlying the decaying systems of the universe. Such order was needed for life on Earth to exist. A miniscule change in the parameters governing the big bang event would have caused the universe to be inhospitable to life. In fact, the universe would not have existed at all.

For example, the amount of matter in the universe is critical in determining the speed of the universe's expansion, which in turn affects its formation. Too little matter, and fast expansion prevents star systems from forming. Too much matter would cause slower expansion and drag the process backward into a "big crunch." We could also go on to discuss, among other things, how the four fundamental forces and the resultant atomic structure were created within precise limits during the big bang (all of these precise limits are part of what was discussed in Chapter 6). The point is that this extreme order proves the big bang does not violate thermodynamics.

Chance could not produce this order that existed at the beginning, so the big bang model has become a thorn to naturalism and integral to intelligent design theory. However, young-earth theorists, who are also intelligent design supporters, ignore this evidence (which is perhaps the strongest evidence in intelligent design theory) simply because of its integral relationship with the universe's antiquity. More on this later as we address the positions and theories encompassed by young-earthism and naturalism starting in Chapter 12.

The big bang definition could also be stated as: "The creation event that occurred at a finite time in the past, which produced the universe (matter, energy and spacetime) out of nothing." Sounds a lot like Genesis 1:1, which is why many secular scientists were against this theory in its early years. Some still question the theory on philosophical rather than scientific grounds. Previously it had been popular to theorize that the universe was infinite in age. This would be necessary in order for a chance-driven universe to have even the slightest hope of being a reality. However, a universe that is infinite in age would violate entropy. If the universe had no beginning, the energy would have dissipated long ago and we would not be here.

Many theories have been proposed in an attempt to unseat the big bang model with most being the result of philosophical objections. None have withstood scrutiny. The big bang validated what theism, especially Christianity, had said all along about the universe having a beginning and starting out of nothing. "Nothing" would be defined as the lack of matter, energy and spacetime. "Something" else logically would have had to existed in order to start the universe. It is also interesting to recall that it was Georges Lemaître, a Belgian priest and astronomer, who first formulated the "hot" big bang model. Big bang skeptic, astronomer Sir Fred Hoyle, named the model "big bang" to make fun of it. Not because of scientific reasons, but for philosophical ones. All this notwithstanding, it is very easy to study the scientific literature and realize that the big bang has considerable scientific evidence supporting it.

In 1929, Edwin Hubble discovered that that the universe was expanding outward in every direction. Interpolating backwards, one could logically surmise that the universe began at a single point.

The release or creation of energy that the explosion-like big bang caused would have been so massive, it was theorized that remnants would still be found eons later. This "cosmic background radiation" was found as predicted, but it was no easy task. Originally these faint signatures were found by accident, to be precisely studied years later by sensitive satellite observations.

When Einstein developed general relativity he found that it predicted an expanding universe, which would make it a prime supporter for what was later known as the big bang. Since the idea of a static, infinite universe was so prevalent at the time, he introduced a "cosmological constant" into the general relativity equations to cancel the effects of the expansion. This made the equations agree with the ideas of the day and is a perfect example of how naturalistic thinking and bad scientific method can be used to rationalize science to support particular conclusions. Einstein would later call the cosmological constant "the greatest blunder of his scientific career.[1]"

In the young-earth article discussed below, they call the cosmic background radiation evidence "very shaky." They do not explain or quote any sources, but by making this claim it is evident they have not actually studied the science or are ignoring it. Dozens of evidences could be cited that substantiate the big bang model. The Notes for this chapter at the end of the book list resources that detail these evidences at great length. The point here is to simply establish that the big bang is a well-supported model and not some wild speculation.

Those creationists, who are critical of the big bang, often point to the debates about the model or continuing refinements. All theories, laws and models are constantly refined as our understanding and ability to test them improves. If this process invalidates a model or law, then by that standard we cannot trust any scientific principles. Here critics are essentially applying a relativistic view on a particular model they do not like for philosophical reasons. The framework of the big bang model is very solid. Certain details, especially those very close to or at the creation event, are what scientists are trying to determine. Various competing theories exist for those details which will be tested by new observations and continuously improving technology.

More than once I have seen claims of big bang invalidity made by quoting a big bang critic who, again, dislikes it because of philosophical reasons, not scientific ones. One example comes from an issue of the *Answers Update* newsletter from the young-earth group Answers in Genesis. They state, "Because Biblical creationists reject the prevailing view among evolutionary astronomers that the universe began with a 'big bang,' his passing [the death of big bang critic Sir Fred Hoyle] is of note to Bible-believing creationists who recognize that the Bible expressly teaches against the big bang.[2]"

This statement is absurd in a number of senses. One, for reasons already described above, a large number of biblical creationists, scholars and scientists in general, agree with the validity of the big bang creation event. Some young-earth creationists like to focus on one or two people who oppose their beliefs, making it sound as if these critics are in the minority. Also notice that the implication is made that if you believe the big bang model to be correct, you are not "Bible-believing." There is more emotion loaded into that statement than science.

> **By the Way:** A creationist is anyone who believes God had anything to do with creating the universe or guiding it along the way. This covers quite a few different positions (dozens, in fact), some of which we will discuss in Chapter 12.

As for "the Bible expressly" teaching against the big bang, I refer you back to how the big bang is actually defined (and its agreement with Genesis). The article footnotes that Genesis "clearly states that the Earth came before the Sun," hence the big bang is wrong. How can Earth exist before the Sun and how can life exist before the Sun? The fact is that Genesis indicates that the light of the Sun was falling upon Earth immediately after its creation. Some people miss this if they do not consider the switch in perspective of the writer from Genesis 1:1 to Genesis 1:2 (see Chapter 14 for more on this "Earth before the Sun" fallacy). The "Earth before the Sun" interpretation of Genesis is illogical and is an invitation to skeptics to ridicule the Bible.

One person commented on Answer in Genesis' article this way: "Of course, what AIG fails to mention is the reason Hoyle was so opposed to the big bang, which was that he could not stomach the idea that the universe had a beginning, because of its theistic/creationist implications. He was so repelled by the idea of a beginning that he proposed an alternative theory that flagrantly violated the First Law of Thermodynamics (basically, continuous creation with no creator).[3]" In other words, Hoyle was distorting science in order to support his chosen worldview (and yet, in spite of his opposition to the big bang, he would admit that, "A commonsense interpretation of the facts suggests that a super intellect has monkeyed with physics, as well as chemistry and biology, and there are no blind forces worth speaking about in nature.[4]"). We will see later how naturalism often distorts science through the abuse of the theory of evolution. In any case, the irony of a creationist group using a skeptic to support its own views should be apparent.

> **Science Bulletin:** The First Law of Thermodynamics states that energy or matter cannot be destroyed. It can, however, change form. Recall the $e=mc^2$ equation. Energy can be converted into matter and vice versa, but neither can be completely destroyed. They can revert to states that are difficult to utilize. That is where the Second Law of Thermodynamics comes into play, which

was mentioned a few paragraphs ago. This law describes how energy seeks a state of equilibrium in all but idealized hypothetical systems.

Some young-earth creationists seem to readily discredit sound scholarship without much thought if that scholarship can be used to support an old-earth in any way. They tack on terms such as "evolutionary" to the big bang or refer to the astronomers as such. This term changing distorts the science and facts. This confuses people because there are actual terms such as stellar evolution, which have nothing to do with naturalistic evolution. It also implies that if you do not agree with the young-earth view, you are automatically compromising to adhere to another view. These methods ignore the fact that naturalistic origins draws nothing from the big bang which is a model of physics and astronomy (more specifically, astrophysics). As already pointed out, the big bang's theistic implications are despised by naturalists (Confusing Term Clarification: *Naturalists* are those who believe in naturalism, not naturalists in the sense of those who study nature).

When young-earthers talk about intelligent design elsewhere, they conveniently leave out the big bang and the fact that intelligent design theory is based on the same science that shows the universe is ancient. They have given no comprehensive response to those who disagree with them or why the many evidences for the big bang are supposedly incorrect. One would think if they had evidence against such an integral part of modern physics that they would be publishing it in the scientific journals. They have not done so. The problem with a lot of young-earth scholarship on this topic is that they will report any theory that declares the "downfall" of the big bang without examining the theory closely. When examined, these theories have always been found to be wild speculation, based on bad physics or an attempt to support personal beliefs (ironically, antireligious beliefs). In other words, bad pseudoscience all around.

It is not the intent here to single out young-earth creationists and make them out to be the enemy of science. However, they do represent a major view of misunderstanding of the big bang in Christianity. Later chapters will elaborate more on both the young-earth and naturalistic viewpoints. Consider for now that a multi-billion year old universe is *young* when compared to the eternal one wanted (and needed) by naturalists. Remember our discussion on the constraints needed for life to exist in the universe? The age of the universe is one such constraint. Earth exists at the *right* time in which the *right* conditions and materials for life exist. Earlier in the history of the universe and the Sun's properties would not be so friendly to life, especially human. A little later in time and similar problems would exist. *The universe is structured in such a way that life is constrained to exist in a particular narrow time frame. We are now living in that time frame.* In fact, it seems that each era on Earth has had "just the right" type of life for the conditions of the day. Chance cannot produce such precision, not even in 14 billion years.

Perhaps both of these groups, instead of ignoring solid scientific evidence and telling science what it must say, should change their *interpretation*. They are both force-fitting their philosophical beliefs onto science and then interpreting the Bible through this "science." A little critical thought will prevent the propagation of errant science, illogical philosophy and misguided theology. At various points, science and religion address the same questions, such as those raised over ultimate origins and purpose. We are capable of honestly following the facts and discoveries of science to the conclusions they construct in spite of the beliefs we bring to the table. When we force preconceived beliefs upon science or the Bible, we are practicing the poorest of scholarship.

11

Floods of Belief

One of the most well known accounts in the Bible is that of Noah and the flood. Many cultures all over the world have similar accounts of a major devastating flood. This seems to be indicative that early man experienced a flood event of immense magnitude.

Many Christians interpret the flood account as being "global" as in the waters covering the entire surface of the planet. Skeptics point to this global flood of Noah as a reason not to trust the Bible because no consistent evidences exist for such an Earth-covering flood. Their belief, however, is contingent on whether or not the global flood interpretation is true.

The global interpretation is not the exclusive interpretation of the Noah account. A lot of people hold to the global view simply because that is what they have been taught since they were young. There is a significant opinion that insists that the flood was local, meaning it was limited to Mesopotamia. Which interpretation is consistent with what the Bible states and agrees with what science has found?

First, it is important to note that the global flood viewpoint did not achieve its current popularity until the second half of the Twentieth Century. This came after much popularization by George McCready Price who was out to disprove Darwinian evolution (the naturalistic version of origins). He believed he could explain geologic formations (which he erroneously believed to be part of naturalistic evolution) by a global flood. He was also intent on keeping with the teachings and "visions" of Seventh Day Adventist "prophetess" Ellen G. White, who preached a global flood and 24-hour creation days. Henry M. Morris and John C. Whitcomb's book *The Genesis Flood* brought Price's ideas further into the mainstream in the 1960s, because they seemed to have the respectable science background that McCready largely lacked.

Prior to this, a local flood and day-age theories had been more widespread in Christianity. They are now becoming dominant again as Christians realize that old age and geology do nothing to help naturalism and that the global flood interpretation has serious problems. First, we need to look at how misused geology fits into the picture.

The "geologic column," or geologic layers, are often used by naturalists as a picture of Darwinian evolution. Supposedly, the layers detail the millions of years of evolution of all species from a common ancestor. This claim fails because the geologic layers show largely distinct divisions with distinct species in each that appear suddenly and in whole. The geologic layers do not show species changing or in transition to other species, as Charles Darwin and others predicted would be found, or would have to be found to support common descent. These issues with naturalistic origins will be discussed in more detail in the next two chapters.

"Deluge Science" is largely a flawed knee-jerk response to this naturalistic science. The global flood is most often posited by young-earth creationists trying to explain away these geologic layers for two reasons. First, they see the layers as "evidence" used by naturalists as described above. Second, such layers require millions of years to form, which violates the young-earth interpretation of Genesis.

The problem with attributing the layers to Noah's flood is obvious to many. How does a raging, global flood produce fairly neat layers? These layers have distinctly different species in each. Examples in the layers exhibiting a chaotic collection of species are rare. By studying the context of the surrounding layers and landscape, these rarities are found to be caused by tectonic movement, local flooding or other localized catastrophes. These localized, spotty "evidences" obviously do not constitute proof of a global flood, yet they are held up as such.

The views of young-earthism will be discussed more in following chapters. But even if geology does not support a global flood, why do many people believe that the Bible clearly teaches a global flood? Even reading the account lightly reveals some problems reconciling the global flood theory with what is written. Perhaps many are reading *too superficially* and relying too much on what *others have told them* the Bible states. Let us look at some reasons why many believe the local flood interpretation is the only consistent and literal reading of the Bible.

1. Genesis 7:11–12 and Genesis 8 clearly state where the floodwaters came from (earthly sources including the atmosphere) and where they returned (into Earth). The water content on Earth today, even considering water vapor loss to space since the flood, is no where near the amount needed for a global flood.

2. The source of water is not enough for a global flood so some people claim Genesis 1:6–7 refers to a "canopy" of water or ice surrounding Earth. To most reading this verse, it is referring to the oceans and the clouds in the sky and the context and original Hebrew do not give any indications of the verse meaning something else. A canopy, its violation of the law of gravity notwithstanding, would drastically change Earth's climate, either heating or cooling it. We see no such climate change in ancient history or the fossil record. This canopy is also claimed to have promoted the long life spans of the day by blocking radiation. The problem with this theory is that water or ice would not stop life threatening cosmic radiation. The canopy theory raises more questions than problems it answers in an attempt to force-fit a global flood interpretation onto Genesis, which has led to its fall from favor among most young earth creationists.

3. In Genesis 7:19–20 we see that all "the high mountains…were covered." The Hebrew for "high mountains" can be literally translated as hills or hill country. The words for "covered" can be translated as "falling upon," "running over" or "residing upon." Another possibility considers Noah's perspective. Floating along on this massive flood, his line of sight would only be a few miles out. To him, everything could have seemed covered as was written. We should also make a note here about ancient Hebrew. Its vocabulary was much smaller than modern Hebrew or English. Many words had multiple meanings whereas we might have a separate word for each meaning. This is why context is often so important.

4. The flood account refers to "the earth" which may seem like it is referring to the entire planet. There is another usage in which "earth" can be literally translated to refer to a particular region. Ancient humanity was believed to be limited to Mesopotamia (we are going way back here), so a local flood would still be "universal" as far as the people alive at that time were concerned. 2 Peter 3:5–6 discusses the flood and its effect on the "…world *at that time*…[my emphasis]" There are numerous other examples of similar usage including Genesis 41:56–57 and 1 Kings 10:24. Ancient mankind was not aware of the existence of most of the world, so what was known was considered the entire Earth. If we were to translate "earth" as being the entire planet, then perhaps Genesis 8:14 would indicate that the planet had become a desert when it states "…the earth was completely dry." Also consider that the Hebrew word that *always* refers to the *entire world* is not used

in the flood account. Only words that can refer to particular regions or peoples are used.

5. The ark did not land on Mt. Ararat as many think. Genesis 8:4 states it landed in the mountains of Ararat. Therefore the ark could have landed anywhere in this region, including the foothills or bases of the larger mountains. Also consider this: Is it not odd that in a global flood that the ark landed only a few miles from where it started?

6. A comparison of the pre-flood Genesis chapters to the post-flood chapters do not show the massive geological changes that a global flood would have caused. For example, the landscape has not changed at all. Noah did not seem lost. Rivers mentioned before the flood remained unchanged which would be kind of odd in a global flood that supposedly created geologic layers in only forty days.

7. Global proponents claim the flood created organic deposits in order to avoid the accepted view that they took millions of years to form. Genesis 6:14 indicates that *possibly* petroleum products were already available before the flood, which are created by eons of decaying organics. I say *possibly*, because the "pitch" in Genesis 6:14 may or may not have been oil-based (but consider Noah would have needed a lot of pitch and he did live in petroleum rich lands). Nevertheless, materials created from dead organisms (oil, coal, limestone, kerogen, marble, topsoil) exist in far greater quantity than a global flood could produce. By no stretch of the imagination was there enough life on Earth, dead or alive, at Noah's time to create all of those materials even if there was a global flood.

8. All the species in the world could not have come from those on the ark without invoking rapid, macroevolution (defined in the next chapter). In fact, the only way young-earth creationists can account for all species being on the ark is by directly or implicitly requiring such a rapid evolution of animals that even naturalists do not subscribe to such a process. The fact is that the Hebrew is particular in the limited types of animals that were brought on the ark. This avoids the need to fit all of the animals of the world onto the ark. Also consider that the precedent set in the Bible concerning punishment always limits it to the intended people and their immediate surroundings.[1] Thus a local flood would only require the destruction of animals closely related to man, i.e. the ones in the area where mankind lived.

9. In many ways, Psalm 104 parallels the creation account in Genesis 1 including how in early Earth's history the entire globe was covered by water before the continents emerged. Verse 6 reveals this fact centuries before scientists had any knowledge of it. In verse 9 the statement is made that "never again will they [the waters] cover the earth." From the context of the Psalm it would be hard to claim this verse is referring to Noah's flood. Hence, verse 9 seems to be directly contrary to a global flood hypothesis.

10. We already described a number of geologic problems with a global flood. Another concerns young-earth creationists rallying around geologic formations initiated by the Mt. St. Helens eruption citing them as proof that formations can be formed fast. The fact is that geology has never denied some formations occur rapidly. We see evidence of both fast and slow processes. These creationists seem to downplay slow processes and highlight fast ones and claim this as "proof." Such selective evidence hardly constitutes proof of anything.

11. It is obvious that some canyons are created by flooding (this seems especially true on Mars, of all places). Some young-earth creationists try to say that the Grand Canyon is one of these canyons (in spite of geologic evidence) and try to fit this into deluge theories (partly to try to "prove" the canyon is "young"). How could a flood carve out a meandering canyon? Also, how does one explain its distinct rock layers?

There are other points concerning the flood that could be discussed, but through this brief discussion it seems that the most literal and consistent interpretation of the flood account is that of the local flood. The local flood is consistent with geology and contains none of the glaring problems and contradictions of the global flood view. It seems that a global flood has been fit onto the Bible simply to support certain young-earth beliefs. This in turn feeds the "Bible is anti-science" perceptions in society because skeptics hold up the global flood as a reason why the Bible is not historical. As we have seen, a global flood fails very easily as bad science or bad biblical interpretation. It does so simply because we took the time to study the science and Bible with slightly more rigor than superficial appraisal. In the process the basis for the skeptics' claims has been removed.

12

Reasonable or Skewed Science: The Evolution vs. Creation Debate

We have already seen examples in the last two chapters detailing how naturalism and creationism can distort science and the scientific method. This chapter begins a four-chapter look at another example. Virtually every book concerning science myths or science and religion issues has an obligatory chapter on creationism versus evolution. The framing of the issues by these books and the media as "creationism vs. evolution" and variants such as "science vs. religion" are grossly misleading and oversimplified. Perhaps this is due to the lack of space, time or ignorance of the subject. Whatever the case may be, most media reports on this issue are superficial in content. In order to understand this issue, we need to first step back and better define these and other viewpoints and exactly what the debate is all about.

The theory of evolution is at the center of the debate. An editorial I read once stated that evolution is not disputed and yet there are sharp disagreements over specifics. If we do not take this as the contradiction it sounds like, we must first define what is meant by evolution. The ability of species to adapt to environmental changes or other stimuli, and perhaps creating what is considered a different species in the process, is not disputed. These different species are related, but rarely, if ever, intermix. It is not uncommon for some adaptations to disappear when the stimulus is removed. The ability for species to change seems written into their genetic code, which is activated as needed. No one disputes this process which is often referred to as microevolution.

Confusing these discussions is that the definition of the term *species* can vary with each usage. There are dozens of definitions to what exactly a species is or what the process of speciation entails. You may hear people refer to humans and cats as different species of life. Technically, this is incorrect. In most classifications of life forms, which are referred to as taxonomy, species is the final division (though *subspecies* is sometimes used). A mountain lion and your housecat are different species of the feline *family*, which is in turn part of the carnivore *order*. Mankind is classified as part of the hominoid family, a division of the primate order. Depending on who is doing the labeling, there may be fourteen levels in the classification process. Here is a list of the most common taxonomic classifications used:

Kingdom > Subkingdom > Phylum > Subphylum > Superclass > Class > Subclass > Infraclass > Order > Superfamily > Family > Genus > Species > Subspecies

The point is that species is often used loosely or defined in ways to fit a particular author's argumentation. In general, two similar animals (such as the two types of cats mentioned above) are usually considered directly related through common ancestors. At least this is what many commonly believe or used to suggest. We will discuss in the next chapter that species with visible similarities (homological classification) do not always turn out to be related.

Returning to microevolution, if environmental influences cause a group of animals to branch off from another, they may be labeled as a new species. This label may result from new geographical location or minor physical changes. They may be virtually identical to their relatives and may or may not be capable of breeding with them. Some people may not want to define them as a new species at all. In other words, microevolution exists, but its results are not always well defined.

These changes in a species, or the origin of a new species (whatever that may be), can be caused by natural selection and genetic mutations. Natural selection can simply be the previously mentioned environmental influence (such as a predator or famine) that wipes out most of a species. Those who survive usually have traits that helped them survive. Some of the traits may have been in their genetic code (recall that these traits may already be active in the animal or activated by environmental influences) or created by random genetic mutations. However, because of the redundancy of the genetic code, the vast majority of mutations are considered "neutral," having no effect—good or bad—upon the affected organism. Mutations do not have the creative power that some naturalists attribute to them. Noticeable or visible mutations, such as birth defects, often kill the animal or decrease its survivability, yet this is the type of mutation necessary for the radical changes naturalists hope for.

These processes are observable on the "micro" level, but can such changes be cumulative over time and create entirely new forms of life such as a mammal from a reptile? This process is usually referred to as macroevolution. Some call *this* speciation (differing from the microevolution variety of speciation) or megaevolution. This macroevolutionary speciation is also used to describe common descent of all life from one organism. Macroevolution is what is hotly debated, not just between scientists and "religious folk," but within the scientific community. Why? Because the evidences needed for proof, and those predicted to exist by the theory of evolution, are shaky and inconsistent. A large body of work questioning the claims of the theory exists and continues to grow.

The next chapter will examine macroevolution and objections to it in more detail, but one has to keep in mind that this is first and foremost a scientific debate. Theological implications come into play later, but the theory of evolution is a valid proposal that must be evaluated and tested. There are parts that have shown validity (microevolution) and those that are debated (macroevolution). Many people allow their philosophical or religious beliefs to corrupt this process of scientific inquiry. It is this colored debate that is most often depicted in the media.

On one side we have those who support the theory of evolution in some form (also known as Darwinian evolution, or Neo-Darwinism as the modern form is called). Evolutionary theory is often guided and influenced by naturalistic philosophy. To recap, naturalism claims science can say nothing about God or a designer and that the universe is the result of undirected processes that preclude the need for a creator. We already established the nature of this particular type of logical fallacy. They are stating *a priori* (before the fact) what science may or may not discover, thus rationalizing their way to preconceived conclusions. This is pursued because if a chance-based universe is not valid, they lose the primary support for naturalism. So the theory of evolution has become a hopeful basis for naturalism and its offshoots.

Using these fallacies, scientists that hold to a naturalistic worldview will claim common descent (or macroevolution) is undisputed and not theoretical. They refuse to acknowledge the research of others, including other Darwinists, who detail macroevolution's problems such as explaining sudden species appearance, lack of transitional forms in the fossil record and the complexity of life.

Some scientists have attempted to come up with an alternative theory to gradual evolution called punctuated equilibrium. This theory attempts to explain the sudden appearance of fully formed life in the fossil record. Orthodox naturalists do not like it because it sounds too much like "divine creation." Other critics of punctuated equilibrium state it only pushes the question of animal and plant creation and the origin of their irreducible complexity "out of sight" but not "out of mind" because they do not provide an adequate mechanism for their "punctuated" appearance.

Related to these Darwinian positions are theistic evolutionists who are creationists that do not adhere to naturalism per se, but believe that macroevolution has at least some validity. Either God let everything go after the initial creation or he guided evolution at times throughout the millennia.

Another major creationist belief is that of old-earth creationists who accept the enormous body of data pointing to Earth's antiquity, but do not find validity with macroevolution. They detail that this is the only way to reconcile the Bible and science without the contradictions and problems that other creationist positions cause.

The third major type of creationist, and the one that usually defines the second side of the evolution debate, is the young-earth creationist view. The terms "creationism," "creationist" or "scientific creationism" nearly always refer to them. They attempt to rationalize science to concur with their interpretation of the Bible. This interpretation states that the creation week in Genesis is made up of 24-hour days. Hence, Earth must only be a few thousand years old. This belief became popular as people mistakenly believed that macroevolution needed an ancient universe to work.

We already addressed that chance could not produce the precise parameters needed for Earth's existence in billions of years (it would not be long enough). The age of the universe (which is about 14 billion years old) is one such parameter and is exactly where it needs to be for life—or specifically Earth—to exist. In other words, according to the laws governing the structure of the universe—as they were "formed" or "created" at the moment of the big bang event—this is the age of the universe that Earth's existence necessitates. Even more simply, Earth could *not* exist in a younger universe defined by the laws of physics as *they* exist. The substances needed for life, their correct quantities and locations would *not* exist in a younger universe.

Science and other creationists are adamant that the young-earth view is contrary to hundreds of dating techniques that consistently date the universe as billions of years old. Human history is also *easily* traced further back than the 10,000 years, which is usually the maximum allowed by young-earth creationists. Yet many still cling to a younger date such as the 4004 B.C. origin date for man that was printed for years in Bibles, which convinced many people it was a fact. Again, this is in spite of well-substantiated dates that conclude mankind is nowhere near being so young (see Appendix C), which is why young-earth creationism is often used by skeptics as a reason not to believe in the Bible or Christianity (apparently many of these skeptics assume young-earthism is *the* Christian viewpoint, or perhaps some want *you* to assume that).

The counting of genealogies in the Bible is often used as a support of young-earthism. What is often forgotten is that in Hebrew tradition many generations are unreported. Often only the famous or infamous are counted. A comparison of the biblical genealogies confirms this technique (compare 1 Chronicles 3:10–12 with Mathew 1:8 and Genesis 5 with Genesis 11 and Luke 3).[1] Also, the Hebrew words for "father" and "son" are less specific than modern usage. For example, the Hebrew for "father" could mean "grandfather" or "great-grandfather," (in other words "father" can mean *ancestor*). Young-earth scholars do not deny these gaps, especially before the time of Abraham, out of necessity to make sense of the chronology of events in the Bible.[2] Some accounts, such as those describing Adam, his life and children, only make sense if the verses are compressing time (see Chapter 14). Genesis 10:25's brief, enigmatic reference to Peleg's day as the time "earth was divided" may give us a reference point in establishing the length of the gaps (see Appendix C). Yet the hope of young-earth creationists is still that these gaps would not cause a creation date in excess of 10,000 years, if that old. Skeptics also ignore these issues in claiming dates derived from the Bible are not accurate.

As for Genesis 1, the Hebrew word for day is translated simply as *day*. Read that again. It does *not* explicitly say 24-hour days. The Hebrew word for day can *literally* have different meanings. Long ages and 12-hour days are just as literal as 24-hour days. So one cannot claim that "24-hour days" is the only literal view. One must consider other contextual items as a whole and not decide beforehand what the conclusions will be. More on the young-earth view will be detailed in Chapter 14.

Throughout the history of Christianity the creation-date subject has been debated, but not until the late Twentieth Century have the beliefs of young-earthism and naturalism become so popular and dogmatic. Both the naturalists and the creationists have controlled the debate through emotionalism and propaganda wars. These two groups often replace science and reasonable discussion with their personal beliefs, which corrupts the scientific method. In the process, the issues have been clouded and more reasonable voices have been left unheard.

An alternate theory, held by scientists of a wide variety of belief systems, is the theory of intelligent design. They have made the claim that modern science shows signs of an intelligent designer. For example, complex information is a clear sign of intelligence. Randomness can produce order (but even what we perceive to be randomness in nature turns out to be based upon order—the constants and natural law we discussed), but not complex information such as the content of this chapter. We will discuss the nature of information in more detail later, but this raises questions such as what is the origin of the complex information in DNA? Another illustration we already discussed is that hundreds of physical constraints must be met just for one planet bearing life to exist. The probability of all these constraints being met by chance is equivalent to zero.

The naturalists do not like this theory because it speaks of a "designer" and claim it is disguised creationism. Some young-earth creationists do not like intelligent design because its principal proponents base it on the same science that concludes the universe is ancient. The supporters counter that the theory uses "designer" generically and the universe's old age is supported by hundreds of straightforward techniques. Unlike some popular forms of creationism, intelligent design attempts to produce a scientific theory to be tested. It simply does not say the other theories are wrong because it says so. Rather it posits if reason and logic lead to a designer, then that is what science will conclude.

In other words, intelligent design is a viable scientific theory to be discussed and debated just like evolution (more on intelligent design theory in Chapter 15). They can stand on their own as scientific theories largely unbiased by personal beliefs, but both have specific implications for religion and worldviews. Many would claim these implications are the domain of theologians. To some degree this is true, but if the science ends up leading to certain "theological" conclusions then it does so. One cannot tell the science what to conclude. If it converges with religion on the same subject, who is to say it is not allowed to do so?

This debate always comes to the forefront when government bodies discuss new science standards for public schools. For a long time it was always colored by the two main groups discussed above. The rise of intelligent design in academia has changed this somewhat. They have pushed the debate to center on practicing reasonable science. This hurts those who want to push their beliefs as fact or minimize or eliminate opposing views altogether.

We should make it hard for philosophies to masquerade as science. Students should be taught to keep an open mind and look at all of the avenues, rather being force-fed debated theories as facts. This is how science is supposed to work. It does not hide from controversies. The lack of emphasis on the scientific method and logic in the standards highlights a flaw in our current educational system. If a clear process of inquiry is not taught, how can students test the theories? We cannot just teach theories and not dissect them to test their validity. That is not science, but a path leading to relativism. We would end up teaching statements like, "Some still believe Earth is flat, but we cannot discuss if they are right or wrong, because someone will be offended."

What we can learn from this debate is that our science standards have a serious flaw. The standards for practicing science—or learning to think critically—are not taught. Instead we focus on what subjects may or may not be taught. Perhaps we need to learn how to teach science, rather than telling science what it may or may not conclude about the universe.

13

Evolving Origins

Once the naturalistic distortions are stripped away from the theory of evolution, it can be studied as a scientific theory instead of a philosophy. Why is the theory so hotly debated? The central core of the theory is that all life is descended from the same organism in the distant past. Through "forces" such as natural selection and genetic mutations, life achieved the diversity and sophistication that we see today.

As we have seen, this theory is also referred to as Darwinism, after Charles Darwin who is the best known formulator of the theory of evolution. The synthesis of genetic aspects (such as genetic mutations) with evolution, which were largely unknown in Darwin's time, is often referred to as Neo-Darwinism. Darwin's writings were mostly theoretical in that they constructed a theory that would be proven *if* certain discoveries were made in the future. The problems with the theory of evolution arise from its inability to make these discoveries and to answer evidence that contradicts common descent. Naturalists ignore or minimize these problems in order to use evolution to justify their belief system.

It is beyond the scope of this book to examine evolutionary theory and all its facets exhaustively. There are quite a few resources that examine this subject in all its aspects. Here we will summarize some of the major scientific objections to modern evolution (Neo-Darwinism).

Information

Someone randomly typing on a computer will produce a sequence of letters that constitute *complex* information (complex in the sense that the letters each form a recognizable pattern). But these letters are *unspecified* since they have no meaning as they stand alone. If the typist happens to produce the consecutive letters "I" and "S" in their sequence then they have produced a *specified* piece of information (since it forms the word "IS"), but without the context of other words, it is meaningless *noncomplex* information. The individual letters still have a complex pattern, but no complex meaning. Realize that they were randomly produced and were *required* by the random tying, not put there with intentions of design (more on this important aspect in Chapter 15). Information that is both complex *and* specified (such as the sentences on this page) and *not* required to exist by virtue of natural laws and is referred to as *complex specified information*, or CSI.

A random process produces either complex unspecified information (the random letters) or noncomplex specified information ("IS"), not CSI. It would be better to call these random products *patterns*, not information. Natural laws or random processes cannot originate information, and our examples are not providing us with any meaningful information, only randomly produced patterns—patterns which can be used to transmit information in a designed context. Natural law and its products can only provide the means to transmit information (such as in DNA discussed next) or produce patterns that are ordered. CSI, however, is only produced by intelligence.

Assume for now that we can easily recognize CSI (we will discuss how in Chapter 15). This recognition has profound implications for chance-based evolution because of the contents found in DNA molecules in living organisms. The DNA in a single cell contains volumes and volumes of complex specified information that define every aspect of that organism from its appearance to its resistance to disease. DNA itself is made up of easily identifiable chemicals, but how do such chemicals produce CSI? They cannot originate information, only carry and transfer it. Also consider that DNA has to exist for the complex organism to live and is interconnected to other molecules such as RNA, which must exist at the *same* time. Evolution is unable to explain how such interdependent complex systems just "appeared" on Earth simultaneously when they cannot survive independent of each other.

Nor can the mutations we discussed create information, they virtually always destroy it. Even a mutation that allows bacteria to resist an antibiotic and pass this trait to its descendents does not add new information to the genome. It simply alters the function of particular genes. This is a physical change, not a change in information content. And new information would be necessary for macroevolutionary level speciation to occur.

Early Earth

Many textbooks refer to the 1953 Miller-Urey experiment as evidence that processes on Earth could produce the complex materials needed for life to exist.

Essentially the experiment tried to mimic the atmosphere they thought existed on ancient Earth. Water was heated in the closed apparatus and gases thought to represent those on early Earth were circulated into contact with an electric spark, simulating lightning. The ensuing reaction produced some of the organic compounds needed by organisms. However, most of the reaction products were not those used by life.

There are a couple of other problems with this still-touted experiment. It was later found that its atmosphere mixture was incorrect and did not represent that of early Earth. Scientists now concede that the early atmosphere was not conducive to the creation of organic compounds. The compounds were not exposed to the kind of conditions they would face in nature—conditions that would destroy them. Lastly, the experiment did not show how complex DNA or RNA could be randomly created. Miller-Urey only showed how a person, not nature, could produce (or *guide* the production of) some organic compounds.

This is not the only problem the ancient Earth poses for the evolutionary paradigm. The earliest life forms are known to have existed or "appeared" during and shortly after periods of significant turmoil in Earth's history. Bombardment from asteroids and comets produced global upheavals, poisoned the atmosphere and caused incredible volcanic activity. How could life have come into being by chance through such an environment hostile to life? Evolutionary theory does not have an explanation.

Appearance of Complex Life

The timing of life's appearance is not the only problem posed to evolution. Its complexity is a larger issue. Evolution paints the development of life going from the very simple (such as single-celled organisms or bacteria) to the most complex (such as humans). The fossil record shows complex life existing early on. Evolutionists will overlook these problems by stating that evolution never claims that an increase in complexity is necessitated by its processes. This is absurd because the increase of complexity among life is what evolution is all about. If this does not happen, then what is evolution trying to explain?

During a geological brief period of time known as the Cambrian explosion about 543 million years ago, as many as a billion new species of life appeared on Earth. Since "species" is such a loose term, it is more exact to say that more than 70 phyla of life came into being. In the time since the Cambrian explosion, no new phyla have appeared. Sudden and fast, with no transitional forms (more on this shortly), the vast majority of life appeared on Earth in the same period of time. Evolutionists can only explain this "biological big bang" by appealing to the nonexistent. They will appeal to fossils yet to be found or some process that may have wiped out evidence supporting the evolutionary paradigm. Some will claim that certain tiny fossils found have filled in the gap, yet these still do not explain how the jump was made shortly thereafter to more complex organisms, and many more of them. Based on the overwhelming evidence that *does exist*, one can reasonably conclude (as most scientists do) that the fossil record is fairly complete, accurate and unscathed. That record shows an explosion of life with little influence from natural causes.

In what seems like a strange throwback to science fiction, some scientists are suggesting the old theory of panspermia. This theory claims Earth was "seeded" for life from space by other intelligences or by accident. The problems with this can be summed up as follows: 1. We have already discussed the improbability of life in the universe, especially complex, intelligent life; and 2. Viable life arriving here from space is extremely unlikely, if not impossible, due to hostile conditions (radiation being the prime problem); and 3. Most importantly, panspermia avoids questions of ultimate origins by moving the location to some distant place. In other words, the theory does not even address the question it claims to answer.

Narrow Window Necessary for Life

This item was the subject of Chapter 6. To review, the constraints that have to be met before life can exist are utterly improbable to have been produced by chance or evolution. This is perhaps the strongest evidence against a chance-driven universe. Some naturalists will now claim that these constraints in some way prove evolution. One supporter wrote "The ID creationist community has adopted the fundamental constants as additional evidence for their Designer of Life—apparently not realizing that many fine-tuning arguments are based on physical constants allowing evolution to proceed.[1]" How do the constants allow this? The author does not explain his circular argument, which essentially states, "Evolution is right, so the physical constants must be what allow evolution to work." His beliefs are contingent on whether or not evolution proceeds as claimed. The fact is that the constants allow life to *exist*, but are so precise and interconnected that chance-based theories could not produce the constants themselves.

These constants that allow life to exist are the foundation of the anthropic principle. This age-old belief, which states that the universe seems tailor-made for life, is the cornerstone of intelligent design theory. It is not odd to be reading articles and books by nontheistic scientists and find them making statements to the effect that they understand why people see so much evidence for the anthropic principle. Headlines on astronomy discoveries in the past twenty years are liberally sprinkled with theological-sounding terms such as finding the "fingerprints of the creator."

These scientists, however, do not always concede to the conclusion of a designer. Part of the problem seems to arise from the belief that anything related to a designer is to be relegated to religion, no matter what science finds. To avoid religious implications, they sometimes come up with relativistic versions of the anthropic principle including ones that explain design by stating if we did not see it, it would not exist. Luckily, few appeal to such absurd notions. Even in the face of strong evidence, it is hard for many people to give up long held beliefs. Often, neat sounding attempts to explain away design are based on flawed reasoning.

Lee Smolin writes in *Three Roads to Quantum Gravity*[2] that the universe does seem to be designed for life. He formulates the anthropic principle as a question and asks, "...why is it that the laws of nature are such that the parameters fall within the narrow ranges needed for life?" He states that concluding "God" as the answer is only valid if "there is no [other] way to explain" the laws of nature. Smolin writes that the God answer is a "God of the Gaps" argument, meaning that God is interjected where we do not have any empirical evidences.

His answers for the anthropic question, however, are not empirical. He admits to what amounts to an a priori assumption that there is nothing outside of the universe. Then he speculates that maybe there is something else, as postulated in multiple universe theories. These theories suggest that many universes exist and at least one was bound to have life. By definition, since we cannot observe such universes, or even indirectly detect them, such theories are unprovable speculation. The design argument (see Chapter 15 for more) does not fall into the same speculative category, so it seems Smolin's reasoning is a "Science of the Gaps" argument.

In *Supersymmetry: Unveiling the Ultimate Laws of the Universe*[3], Gordon Kane suggests that if the dinosaurs had not been wiped out by an asteroid, they would be here claiming the universe was designed for them. In other words, the anthropic principle should apply to humans at any time in Earth's history, if it applies to us at all. The primary problem with this reasoning is that one cannot base a theory on an alternative history. The dinosaurs did not survive, so to pretend what would have happened if they had made it may be interesting speculation, but it is not reality. Science is based on fact, not fiction. Also, consider that the dinosaurs seemed to have been wiped out at the right time in history, by the right size asteroid that hit the right location to ready Earth for the arrival new life. Any deviations and we would not be here. Coincidences or design?

Kane also implies that design is only apparent because, "…a century and a half of study has yielded extensive documentation to confirm evolution of the human eye and brain and body." This chapter explains how this definitely is *not* the case, especially when concerning humans. It is not difficult to find in the biological literature the uncertainty coming from naturalists and Darwinists. Yet, some writers like Kane expect us to forget all of that with one sure-sounding sentence. Perhaps the over-specialization of fields explains why such comments are common in popular astronomy or physics books. These scientists should once and awhile cross over and actually study what the state of other sciences are since they all are ultimately interconnected.

Carl Sagan perpetually held out hope for the discovery of extraterrestrial life to verify his naturalistic worldview. In his book *Pale Blue Dot*[4] he continued this thinking, while trying his best to debunk the anthropic principle. His reasoning, however, was very thin. For example, he admits that "certain" laws and constants seem consistent for "our kind" of life. Then he comments, "But essentially the same laws and constants are required to make a rock." Not exactly. The key word is *essentially*, which means here *not entirely*.

A change in the laws may have produced a universe with only rocks, but that would not explain how such fine-tuned laws for rocks came into being. As with Kane's argument, Sagan tries to replace what we find in the real world (life allowed by narrow rules) with a fictional idea (a rock world). We have to explain what exists and why, not what could have been if things were different. Obviously, they were not different, which is the point: Why is the universe *not* different from what we find?

Transitional Forms

Another major problem for evolution is the lack of transitional forms in the fossil record. It has long been predicted that fossils should reveal many organisms "in transition" between different types. What the record does reveal is a history of mass extinctions and sudden appearances of new complex types. After each extinction (brought about by various mechanisms such as impact events), hundreds and sometimes thousands of life forms appear in their final form without transitions.

Some fossils are claimed to be transitional, but they are few and far between and hotly debated. Millions of year's worth of fossil layers should produce consistent and numerous transitional fossils if evolutionary theory is correct. The fossils do not produce such evidences. The most popular transitional debate is over the possible link between birds and dinosaurs. Dinosaur fossils that are similar to birds, or birds that seem similar to dinosaurs, are held up as transitionals. There are a couple problems with this debate that are usually glossed over.

First, defining these, or any, fossils as transitionals is based largely upon appearance. Determining the relation of various life forms used to be chiefly based on these appearances, also known as homology. However, we can find similar features between many very different animal types. Exactly what is the duck-billed platypus a transition from or to? In fact, many types of animals originally thought to be related or descended from the same ancestor—such as two types of river dolphins which look virtually identical—have been shown through genetics to have developed independently. Genetics is revealing that homology is often a poor indicator of relation.

Another point of contention over dinosaur-to-bird fossils is that in spite of similarities between certain types of birds and dinosaurs, their differences represent an impassable gulf. This gulf is known as biochemical complexity or irreducible complexity. The biochemical systems of any organism are extremely complex and interdependent. Remove or damage one system, many others are affected and the organism will die or have a greatly reduced life span. The origin of new, complex biochemical systems, such as those needed to create the avian lung or flight itself, cannot be created piecemeal without endangering the organism or killing it.

This is why evolution's idea of cumulative steps producing new traits, which in turn are supposed to produce entirely new life forms, is problematic. Creating such a new form of life requires a complete and simultaneous change of major and minor biochemical systems. Small, singular changes are more apt to be ignored and larger ones seen as defects by the organism. What kind of mechanism could produce the structural changes in a dinosaur to gradually or suddenly allow it to become bird-like? Evolutionists do not know.

Also realize that whether or not these transitionals are indeed transitions is often based on who is defining the fossil. Is this fossil simply a bird-like dinosaur or a dinosaur-like bird? Or is it really a transition? If we were to assume for a moment that these fossils are transitional forms, we still have the serious problem known as the temporal paradox. These supposed transitional forms are in the fossil record *after* the first known, fully formed, undisputed bird fossils. Also consider that because the "transitionals" are *fully formed* in all their components, they are *not* in transition by definition. No *partial* developments indicating a future transformation. The logical conclusion is that we cannot consider these fossils transitions to anything.

In the end, the problems with changing from one complex system to another is the simplest reason of why the fossil record is devoid of undisputed transitions. Consider one last example, the giraffe. We do not find many fossils attesting to the "evolution" or "transition" of the giraffe from earlier ancestors. This does not stop evolutionists from trying to explain its origin. Its long neck and legs were supposedly formed to overcome a need to feed off trees with each generation having slightly longer necks and legs. This height introduces the problem of making it difficult for the giraffe to drink by creating pressure changes in the circulatory system when it bends its neck to reach the ground.

Without an exceedingly complex system to control pressure changes, the brain would hemorrhage and the giraffe would die when it bent over to drink water. This system had to develop simultaneously with the gradual expansion of the neck from generation to generation. Assuming for a moment that each giraffe could indeed pass on its stretched neck (produced by trying to reach higher branches) to the next generation, exactly how would this produce the advanced pressure control system to keep the giraffe alive? Evolution cannot explain the development of this necessary survival feature in the giraffe. The astute observer may also conclude that if giraffes so badly needed to reach trees to survive, they would have died long before they grew long necks.

Repeatable Evolution and Convergent Evolution

Repeatable evolution is the appearance of different animals with similar characteristics that have formed under similar natural selection influences. We addressed this somewhat in discussing transitional forms. An example of repeatable evolution would be the different species of river dolphins, which were thought to have emerged from a common ancestor. DNA analysis reveals they developed independently.

These types of life present a major problem for evolutionary theory. Evolution is rooted in the "forces" of chance, which means *few* life forms should be similar. Those that are alike should be descended from the same ancestor. Genetic studies are showing that many species that seem very similar are unrelated. How is chance producing so many similar, unrelated species?

Closely related to the concept of repeatable evolution is convergent evolution. This is when two widely separated, unrelated organisms produce similar characteristics. Here the non-relation of the animals is obvious. Take for example dolphins and fish. Dolphins are mammals with very different biochemical systems, yet live in the same environment as fish. Birds and bats are another example. Evolutionists explain these similarities as the animals' response to similar environments. However, the commonality of convergence is extremely high for chance-based evolution. In other words, how does chance arrive at similar conclusions for similar problems so often? Add to this the fact that the differences in the complex biochemical systems between dolphins and fish or birds and bats are so vast, chance could not produce one of them let alone hundreds of different complex systems.

More on Homology, The Circular Argument

We have already seen the problems with using similar appearance or structure to infer relation or common descent among life. There is another flaw in this use of homology by evolutionists. Neo-Darwinists claim homology is the result of common descent and *at the same time* state homology is a proof of common descent. This is a classic circular argument that has been pointed out for decades. One cannot claim that item #1 (homology) *arises* from item #2 (common descent) and then say at the same time that item #1 proves item #2 is true.

On a related note, the similar structural patterns (such as bone structure) that taxonomy was (and often still is) based upon, are sometimes pointed to as evidence for an intelligent designer. Common structures, same fundamentals (DNA, carbon-based life, and so forth) and other similarities are seen as a sign of a common designer. As we have noted, the differences that exist are enough to preclude common descent and random causes. DNA was once held up as proof of common descent because it is in all living things. Then researchers began finding significant differences in the DNA that should not be there if common descent is true. Homology, as in this DNA example, fails to support common descent while perhaps at the same time providing an indicator for design.

The Origin of Man

The evolution of man is one the most hotly debated issues in evolutionary theory. Many evolutionists do not understand why many people still do not agree with them that primates and man share the same ancestor. They blame it on religious beliefs, but for many people the first reason is that the theory is presented with such flimsily evidence.

The "family tree" of man and primates has been built using extremely fragmented finds. A partial skull here, a bone fragment there. Many people find it ridiculous that some scientists try to create a detailed history of man from these finds. On top of that, no fossils exist connecting us to the living primates that are supposed to be our relatives. The theory of common descent of primates and man is largely based on some physical similarities, while the significant differences (which we are about to detail) are ignored. Such profound theorizing on scanty evidence is very poor science. At least the controversy over dinosaurs and birds has complete skeletons to debate over.

"But I always hear the claim that we share 98.5% of DNA with certain primates!" That is a very misleading statement that vastly oversimplifies genetics. For example, if someone told you an organism shared 75% of the same genes with humans, you would think we would have significant similarities with that animal. Actually, the organisms in question are the soil dwelling nematodes. They are about four thousandths of an inch long. Genetics is obviously a bit more complicated than the "98.5%" figure suggests. Also, each primate species appears suddenly and morphologically complete. There is no intermediate development and no explanation for how the drastic biochemical differences between each species could have occurred.

Neanderthals (now usually referred to as Neandertals) were often said to be ancestors or relations of mankind, making them a link to more primitive ancestors. Were they human or simply another lesser, unrelated hominid species? Genetics and other studies are pointing to the latter. Naturalists have long wanted Neanderthals to be our relations or be the ancestors of humans to help support the theory of man's descent from ancient primates. Ironically, some creationists have also tried to show a connection between Neanderthals and humans in order to *disprove* common descent. Their reasoning is that if Neanderthals and humans are the same species, then no evolution took place.

These creationists point to Neanderthal use of art, tools and burial as an indication of humanity. This is poor standard because these traits can be seen in many animals, including existing primates. Physical and genetic studies both conclude that Neanderthals were not human, nor did they contribute to our genetic make-up. Physical and morphological differences include no tear ducts, differing design of skull and jawbone and different nasal structure. The conclusion that Neanderthals are not our relatives or ancestors is a dramatic shift in science that is quickly becoming accepted in spite of its implications for naturalistic origins. However, young-earth creationists are still claiming Neanderthals are humans in spite of the convincing evidence to the contrary! Also, watch out for some recent books on origins, which still reference out-of-date research, much as changed in recent years. The Neanderthal issue has made mankind's supposed descent from primates even more fragmentary and questionable.

Further genetic studies have confirmed a recent date for mankind's origin. Some studies put the human species' origin as far back as 150,000 years ago, give or take a couple centuries. Most recent studies refine that date to 60,000–50,000 years ago. New research and improved technologies are continuously refining the dates, but here is how one determines the dates. There is a type of DNA known as mitochondrial DNA (or mtDNA) that is in both men and women. Both receive this mtDNA *only* from their mothers. By comparing the differences in living mtDNA to the mtDNA found in ancient human remains and populations around the world (especially isolated people groups), we can determine the time these mutations took to form. Mutation rates thus serve as a clock to determine when any two individuals had a common ancestor or when mankind originated. A similar method using a gene of the Y-chromosomes that are only passed through men confirms the technique. The use of these methods were also used to confirm that Neanderthals are not related to mankind. Also, similar genetic studies of human diseases confirm mankind's origin date and migration around the globe.

By studying the genetics of the oldest people groups of the world, the 50,000 years ago origin date is derived for a common female ancestor for *all* people. This ancestor has been referred to as the "Mitochondrial Eve." Naturalists will claim this origin date is when humans "broke off" or formed from an older primate species. This raises some questions for the naturalists that they have yet to answer. How did the human species suddenly "appear?" No gradual or intermediate species can be found. How does one explain the significant physical and genetic differences with primates? Such changes do not happen overnight, yet they appear together at once. The human species itself has undergone some changes over the millennia. We live longer (though life-span differences are not as great as often reported), are taller and have greater intelligence than our ancestors of only a few centuries ago. Granted, many of these changes resulted from our own intelligence-induced changes in medical advances and technology. Yet these, and our physical adaptations to different environments on Earth, are minor when compared to outright differences in morphology and genetics with existing primate species.

Most naturalists overlook this problem, at least in print. They assume man did break off from these earlier primates (note that some hominid species, like Neanderthals, overlapped with man's existence). This assumption is based on what the theory of common descent *requires* for it to be valid, not what empirical evidence actually supports. Young-earth creationists are still not happy with the younger dates of mankind's origin because in their theory, they rarely exceed 10,000 years ago as an origin date. The problem with this is that mankind's history can reliably be traced further back than that date. In any case, humans have been around for a long time, but not long enough to support the naturalistic science. Appendix C contains a timeline of mankind's early history to put all of this into perspective.

All of this modern research also throws a monkey wrench into the ideas of people who try to use "race" to discriminate or divide people. Being that all people are descended from the same ancestors, "race" becomes nothing more than an indicator of cultural and geographical origins. Science has shown we are all part of one *human race* descended from the *same* ancestors.

Other items troubling to naturalistic theories are human language and our brain. Language appeared suddenly, fully formed and complex. It has grown in vocabulary, but an increase in size is not the same as an increase in complexity. If anything, languages have become simpler by people choosing primary languages—English is considered the international language—and by sharing words from each other. Our brain's capacity for intelligence and social interaction is unmatched by any animal. No primate species—living or extinct—has or had such capabilities. Once again we have a sudden appearance of a complex organism, with no intermediate stages, with the very first humans.

Human characteristics such as the brain and language (and complex vocal communication) are unique. They entail complexity unproducible by chance. If chance did have such power, why were only humans endowed with such features out of millions of life forms? Surely the force of chance that supposedly produced so many similar life forms against all odds, would create more than one species on par with humans. As we have seen, chance, natural selection, etc., does not have such power. Nor can they explain the world we see around us. Naturalists and evolutionists often still call ancient hominids "human ancestors" in spite of all the issues we have discussed. Perhaps they are hoping the evidence will all disappear. Ultimately, however, human beings are the one living thing on this planet that can make even the most dedicated naturalist restless in their theorizing and the most ardent atheist pause and consider a designer. It almost seems, in a sense, that *we* are the aliens on this world.

Can We Stop Yet?

There are other problems with the theory of evolution that could be cited, volumes have been written on the subject. These have been some of the most significant and should be enough to show that this theory is problematic at best. Many supporters of evolutionary theory will claim these objections come only from those people with a religious ax to grind or are creationist attacks. Notice, however, that these problems are *scientific* objections

One last note on the "religion versus evolution" paradigm: What did Pope John Paul II say a few years ago about evolution? Depending on what you read by some naturalists *and* creationists, it is often claimed he endorsed evolution. Well, no, that is not what he said. He seemed to indicate that evolution is a theory worth studying, however he added some caveats. Phillip E. Johnson explains:

> ...the pope went on to say that some versions of the theory of evolution are not strictly derived from data but also "borrow certain notions from natural philosophy....Hence the existence of materialist, reductionist and spiritualist explanations...[any] Theories of evolution which, in accordance with the philosophies inspiring them, consider the spirit as emerging from the forces of living matter or as a mere epiphenomenon of this matter, are incompatible with the truth about man." [The pope]...drew the line between (1) legitimate scientific theories based on empirical evidence, which the Church will honor, and (2) overly ambitious manifestations of materialist philosophy which contradict truths which are fundamental to the Church's magisterium.[5]

The whole point is that Neo-Darwinism fails on its own as a scientific theory as much as the naturalistic worldview that appeals to it for support. Why is this theory touted as factual so often? As we have seen, people driven by the naturalistic worldview replace good science with their philosophy. They replace empirical evidence with assumptions and preconceived notions. Some seem to hope that these problems will go away or no one will notice. Maybe someone will find a true transitional or can explain why so much life appeared during the Cambrian explosion and why "speciation" at that level has since ceased to exist. Maybe genetics will stop distancing man from primates. There are a lot of "maybes" for those who practice bad science. At some point, as problems with and objections to Neo-Darwinism continue to grow, perhaps these scientists will start practicing good science and follow the evidence to where it leads.

14

The Genesis Question

As Chapter 12 detailed, the second side that has usually defined the evolution vs. creation debate is the position of young-earth creationism. Virtually every book on myths in science and in defense of naturalism has at least one chapter dedicated to creationism. Remember that *creationism, creation science* and *scientific creationism* nearly always refer to the young-earth variety of creationism. The focus of naturalists has recently become more directed to the supporters of intelligent design theory (see the next chapter) since that theory is well defined and appeals to real empirical study. Nevertheless, they continue to stereotype young-earthers as being representative of creationism and Christianity.

We have already discussed how the flood geology component of creationism faces various scientific and biblical problems. You may recall that a global flood is integral to the young-earth creationists' ability to explain the geologic layers. Our discussion on the big bang revealed the penchant for creationists to abandon otherwise sound science because it also shows the universe to be ancient. This arises from the misconception that an ancient universe is vital to the theory of evolution. This belief is propagated indirectly by evolutionists who often defer to vast periods of time as a stopgap measure for things they cannot explain. Biomechanical scientist Neil Broom explains:

> Almost all the origin-of-life hypotheses exploit the notion that, because vast periods of time were available for the evolution of life to occur, any apparent weakness or inadequacy in a particular naturalistic scenario, is more than compensated for by the potential for creative innovation provided by the enormous scale of time...[megatime] appears to have relieved many of the obligation to provide rigorously argued scientific explanations...[this] characteristic explanatory sloppiness...is used as a kind of magic wand...in order to accomplish remarkable feats of materialistic magic.[1]

We have already discussed how millions of years cannot account for the precision of constants that allow life to exist. Chance would need a much longer time to even hope to produce such precision. The "ancient universe equals evolution" equation that many creationists posit is not valid. Now let us look to why the universe is indeed ancient (between 13 and 14 billion years old).

There are hundreds of methods, not just one or two, that are used to date Earth's and the universe's age. Some are simple, some a little more complex, but all boil down to basic physics and mathematics. Consider this illustration that uses basic geometry: As Earth moves around the Sun, certain stars appear to move slightly against the background of more distant stars. This movement is called stellar parallax. This is essentially the same effect that occurs if you were to hold a finger at arms length, then alternately close one eye and then the other. Your finger appears to shift back and forth against the background.

The angle between the two observed positions (six months apart) of the particular star (which is determined by simple geometry with the known distance of Earth from the Sun—93 million miles, one astronomical unit or AU) is plugged into the geometric equation $d = 1/p$, with p being the angle (in seconds). Solving for d, we find the distance of the star (in parsecs) from Earth. One parsec is equal to 206,265 astronomical units.

Using this method to measure the distance to nearby stars that are light-years away, we find that we are seeing stars as they were many *years* ago. Similarly, the Sun is far enough away that its light takes about eight minutes to reach us. We are seeing the Sun as it was eight minutes in the past.

> **We Can Study the Past**: Young-earth creationist Ken Ham has written, "…the science that put man on the moon can't be used to directly observe the past. Scientists don't have the past to study.[2]" As we have seen with the light of the Sun, Ham's statement is untrue. Also consider that even if a fossil were only created *yesterday*, we would still be studying the past. Astronomy, archaeology, geology and paleontology all rely on the ability to study the past. The latter three actually have tangible samples to study. To say we do not have the past to study or cannot directly observe it, is to deny reason and reality.

Recall that the speed of light cannot change without destroying the very structure of the universe. So as we examine distant objects, we quickly realize that we are seeing objects as they were millions of years ago. As we examine more and more distant objects, parallax becomes more difficult to measure, so we must employ additional methods to determine ages (however, satellites have extended the effective range and accuracy of parallax measurements). The objects in space we observe at billions of light-years away are one of the strongest evidences against a young universe. Creationists have completely failed to come up with an alternative explanation that has withstood even simple testing.

The nuclear age has brought with it nuclear physics. The workings of fusion are well known through our production of fusion bombs and the development of fusion test-reactors that may someday provide us with energy. All stars burn by using this fusion process. The observable size of stars (along with other traits) coupled with our knowledge of fusion can be used to determine how long they have been burning. The dates come to billions of years, not hundreds.

There are many other fairly simple measurement techniques that are often used to verify each other. The point here is that determining the age of the universe is not some unknowable, magical process known only to the select few. The fact that these methods are essentially very basic means it would be hard to bamboozle the public at large to believe the world was older than it is in actuality. Young-earth creationists are well aware of this and some will even indicate that the "science" part of their theory is their weakest argument.[3] This is why we will spend little time on individual young-earth "evidences" and instead are focusing on the ease and reality of dating the universe ancient.

Radiometric Dating

The set of dating methods that creationists most often focus on are the radiometric variety. These are based on simpler concepts than their name may imply. Radioactive elements undergo a decay process, which form a new "daughter" atom. For example, uranium-238 (the parent radioactive isotope) decays into lead-206 (the daughter decay product). The number after the dash denotes the particular version, or isotope, of that element (isotopes can also be noted as ^{238}U, etc.). This number is the atomic weight, which is the number of neutrons and protons present. The time it takes for half of the parent atoms of the sample to decay is known as the half-life. Compare the daughter product to the remaining parent isotope and a date can be determined.

In other words, the radiometric dating process is based on some basic, well-understood physics and mathematics, not some mysterious ever-changing process. Consider the following: Aluminum-26 has a half-life of 700,000 years. If the universe is old as it is claimed, we should find none of this isotope left. If the universe is young, there should be plenty of aluminum-26 lying around since it has had little time to decay. What we do find is that a lot of its daughter product magnesium-26 exists. More simply, the quantity of the daughter product—and lack of source aluminum-26—necessitate that a supply of aluminum-26 must have existed more than 700,000 years ago. Some isotopes with short half-lives can still be found because they are being replenished by some natural process (see Carbon-14 discussion next) or are produced by man from other isotopes for various uses. Interestingly, concentrations of aluminum-26 have been found in space suggesting relatively recent events such as supernovas that would have created it (again, levels of magneiusm-26 in the universe show that much more source aluminum-26 *must* have existed in the past). Extinct isotopes, identified by the existing daughter products, are a very strong evidence for the antiquity of the universe (we also know they had to have existed by virtue of the physical processes such as stellar fusion and other events in the stellar life cycle). In fact, *every* element has extinct isotopes. Also note that the oldest half-lives do not exceed the old date for the universe's origin. In other words, radiometric dating provides limits for how old—and how young—the universe can possibly be.

Creationists will often try to make these methods sound unreliable through complex-sounding refutations. These arguments usually focus on known errors or the limitations of a particular dating method.

What the creationist literature neglects to mention is that scientists are aware of each radiometric method's weakness and strength. This is why multiple methods are often used and new, more sophisticated methods are often developed as advances in technology allow. Too often, examples of "unreliable dating methods" concern known mistakes someone made. Improper procedures and safeguards are often the cause of bad datings. Mistakes, however, do not invalidate an entire science. Never are the thousands of unrefuted datings ever discussed. Selective evidence ultimately proves nothing.

Carbon dating is perhaps the most maligned dating method. This method uses carbon-14, which decays into nitrogen-14. The percentage of the original carbon-14 left in a specimen (while keeping its half-life of about 5,730 years in mind) determines its age. While alive, the carbon-14 supply for the specimen is constant (obtained by absorbing carbon dioxide from the atmosphere), but begins to decay once the organism dies. The original amount of Carbon-14 is determined by examining the amount of carbon-12 in the specimen. Carbon-12 does not decay and the ratio between its atoms and that of carbon-14 are always fairly constant. Since the carbon-14 content of an organism is determined by the carbon-14 content in the atmosphere while it is alive—and these atmospheric amounts, while fairly constant, can vary somewhat over time—further calibration is accomplished by studying tree ring datings (dendrochronology) and ice core samples which can reveal carbon-14 content in the atmosphere of the past. This helps to confirm what the ratio of carbon-12 to carbon-14 in the specimen was when it died. Hence, by comparing the amount of Carbon-14 left to how much Carbon-12 is present, and using the half-life calculation, an age for the specimen can be determined.

Carbon dating is useful to date specimens up to about 50,000 years old (this can be extended to about 100,000 years by using technology to count atoms). In the end, consistent and unchallenged datings are performed on many specimens, but this method is *only* useful for relatively *young* and *organic* specimens. It has no use in dating the age of Earth, the universe, etc. Of interesting note to young-earth creationists is that anthracite coal does not contain any Carbon-14. Hence, this substance formed from organics older in age than young-earth creationists admit possible.

Finally, the creationist claim that dating objects is "circular reasoning"—rocks date fossils and the fossils in turn date the rocks—falls apart in light of our analysis. While such a criticism may have been valid before the widespread use of radiometric dating techniques, the claim is now entirely contingent on the accuracy and usefulness of these techniques. As we have seen, these techniques are neither inaccurate nor unknowable.

Bad Creation Science

Some creationists will point out that scientists always seem to be changing the age of the universe. This goes back to previous discussions where the point was raised on how the advancement of technology refines science. We can see further into space now than ten years ago. Our computers can produce and test more sophisticated models. Theories and even laws are having their details refined constantly. Realize that such refinements have removed contradictions such as stars that appeared older than the universe, but have given no evidence for a young universe.

The origin of the universe has been determined to be "...13.7 billion years old, with a remarkably small one percent margin of error[4]" by sensitive space-based observations. This is not far off from previous estimates and it confirms the universe is old, but with better precision. No discovery or research gives any indication that the universe is even remotely close to 10,000 years old (with some creationists believing it is much younger). It is hard to argue with a "one percent margin of error" and archaeology and history also confirm that a young date to be unreasonable because mankind's history can easily be traced further than young-earth creationists allow.

It is also important to reiterate that in spite of claiming to have "proof" for a young universe, such proofs do not appear in the scientific journals. These professional journals are where scientists submit papers on the latest discoveries, experiments, etc., to be reviewed by peers around the world. The reviewing and refinement of such work is open and sometimes heated, but the solid works survive intact or with some modification. If young-earth creationists have proofs that the universe is young—a jaw-dropping, science-changing claim by any standard—why no peer-reviewed articles? The only articles to be found are in their own isolated publications.

The only explanation is that the young-earth scientists would claim that no naturalistic-biased journal would ever print their articles. Granted, bias often creeps into many publications, but beyond that, this is not a valid excuse. The research that has found life rare in the universe and a universe "fine-tuned" for the life that does exist, has been put through the peer review process in spite of its theological implications. So if the universe is young, where is the scientific data that will withstand the unrelenting scrutiny that science places on all theories? It is a very similar question to the ones asked of Neo-Darwinism (Where is the evidence of common descent and chance-produced information?).

Some popular "evidences" for a young-earth were printed for years after they were thoroughly disproved. These marked what seemed to be a pattern of "throwing things out there" to see if they stuck or convinced people who never test what they read or hear. For example, the claim was made that there is not enough dust on the Moon for it to be billions of years old. This was based on old estimates of how much material arrives from space to the Moon and Earth. Even after *direct* measurements in space were made that proved that the dust quantity on the Moon was what it would have received after billions of years, the Moon dust argument was still propagated for about *three decades*.

Young-earth creationists also insisted for just as long that the continental plates did not move, because such a process would take millions of years. In spite of movements caused by earthquakes and the puzzle like appearance of South America and Africa that still exists after eons of erosion, plate-tectonics was a major paradigm shift for geology. Now space-based satellites measure the subtle movements of continents using lasers. Few in science now would try to dispute that the plates move in the face these evidences.

The opposition to tectonics has been dropped by some creationists. There are those who try to attribute the plate movements to their current positions as a result of a global flood. If the flood was so violent to cause such upheaval, how did Noah survive? Divine protection is invoked. Ok, how do you explain that the post-flood world was not so different from the pre-flood world? We discussed this problem before and it is a serious problem for explaining away tectonic movement by a global flood. "Just so" stories are not enough to prove a theory.

These are but two of a series of recent reversals on long-standing young-earth positions. Perhaps this is an attempt to clean up obviously bad scientific scholarship that is held up by skeptics, naturalists and other creationists as reasons not to listen to young-earth supporters. Sometimes these capitulations are presented in carefully couched terminology. One source blames the original Moon dust datings on "evolutionists" and concludes that the new data does not prove either a young or old Moon.[5] How can it not be either? The dust layer matches the observed material influx rate for an old Moon, not a young one.

Even if this is an attempt to improve scholarship and image, one can surmise one conclusion. It seems that in light of overwhelming evidence for the universe's age and the validity of such science, that a domino effect has begun. Young-earth "evidences" are becoming harder and harder to defend (for one last comment on young-earth science, see The One-Sided Equation in Appendix D). Creationists are now relying more on biblical arguments than scientific ones. That is where we will now focus our attention.

Does the Bible Really Teach a Young-Earth?

Listening to some creationists, one would get the impression that belief in a young-earth must be fundamental to Christian beliefs. In fact, some have formulated this belief to be critical to the fundamental Christian doctrine of salvation. This belief could be paraphrased as "If you don't belief in a *literal* seven-day Genesis then you undermine the accuracy of the Bible and fundamental doctrines such as salvation." The validity of this statement is entirely contingent on the word *literal*. As we discussed in a previous chapter, the meaning of the Hebrew word day in Genesis 1 can be *literally* translated more than one way. So one must look at the various contextual elements of Genesis 1 before declaring what is the literal truth and what is not.

Not all creationists make their interpretation a "matter of orthodoxy" or "test of salvation." Virtually none of the denominations in the orthodoxy of Christianity have declared a seven-day creation the preferred translation, and none have claimed it is pivotal to belief or salvation. Both sides can quote church fathers all they want who supported one or the other side or debate what these people believed was true. Listing these noteworthy people seems to constitute proof to many people, especially on the young-earth side. However, they are ignoring that modern scholars have better translations and superior understanding of the original languages, cultures and *especially* science. More importantly they are missing the point that one would be hard pressed to find many (if any) of these founders or scholars that believed the length of creation is essential to salvation. Picking and choosing ancient, long-deceased people who may or may not have supported your *current* theory does not constitute proof. The ability of your theory to be reasonably constructed, to provide evidences and to withstand testing is what matters.

Let us now turn to specific biblical problems for the young-earth paradigm. Young-earth creationists state that a *simple* reading of Genesis 1 and 2 clearly indicates the creation week is made up of 24-hour days. Is this simple reading bypassing careful study of the Bible and replacing it with a *superficial* reading? One can read the Bible "simply" and have significant understanding of its contents. "Simple," though, does not preclude engaging one's brain and not thinking about what is being read. The argument can be made that a simple reading of Genesis 1 and 2 *does* provide clear evidence that the creation week was *not* made up of 24-hour days.

1. Genesis 1:3 has light appearing a few days before the Sun is mentioned in Genesis 1:16. Is it rational for God (a rational being) to have light existing before the sources are created? Does it really make sense to think life existed before the Sun formed? Or that Earth could exist without the Sun? Consider that starting with verse Genesis 1:2, the account is written from the perspective of what someone on the planet's surface would have seen. In other words, he his writing what would have been seen from Earth if someone had been there at that time (and all creationists would probably agree that no one was actually around during these creation "days"). Light *existed* in Genesis 1:3, but the Sun and stars were not *seen* until later. This matches precisely with scientific evidence that tells of a cloudy early Earth, with the Sun only visible later (these clouds would not be the "canopy" discussed in Chapter 11. The sky is already clear in Genesis 1:14 while canopy proponents require it to exist until Noah's day). Chapter 1:16 writes "God made two great lights..." which some scholars believe is better translated as "God *had* made." "Made" is translated from a Hebrew word that would seem to indicate the light coming from something created previously.

2. Genesis 1 does not refer to the "days" as 24-hour days. The text only reads as *day*, so you *have* to look at the context. The New International Version (NIV) and some other translations set the days off differently, and more

accurate to the Hebrew, than do other translations. The King James Version (KJV), or ones that over-simplify such as The Living Bible (TLB), are not as accurate to the Hebrew and make it sound as if these were 24-hour days. Compare and you will see the difference. KJV: "And the evening and morning were the first day." NIV: "And there was evening, and there was morning—the first day." The Hebrew matches the latter translation more precisely, which shows that a 24-hour day is not as obvious as some claim. If it were a 24-hour day, one would expect it to *obviously* say so. The text, however, seems to be indicating something else.

3. The attaching of an ordinal (such as "first") or other appendage (such as "long") to day does not always indicate a 24-hour day. See Zechariah 14:7, which uses "one day" or "a day" depending on the translation and Hosea 6:2. Scholars have long interpreted the use of day in these prophetic verses as meaning years or longer periods. There is no good reason to dismiss these examples simply because they are considered prophecy. In 1 Samuel 7:2, the word for day is translated as "long time" or "the time was long" and refers to twenty years. In Deuteronomy 10:10, day is translated as "the first time" and refers to forty days. In 1 Chronicles 29:27 the word for day is translated as "the time" and refers to forty years (some translations leave it out since the context makes it repetitive).

4. Similarly, the Hebrew for the phrase "evening and morning" or "evening, and there was morning" has usages not limited to 24-hour days. In fact, there are numerous usages in the Bible that this phrase, or variants of it, refer to continuous processes or activities. Exodus 18:13, 27:21, Leviticus 24:2–3 and Daniel 8:14,26 all use this phrase in a context of something that occurs on a continual basis over more than one 24-hour day.

5. The third day must have been longer than 24-hours, since the text indicates a process that would take a year or longer. On this day, the text specifically states that the land produced plants and trees. After they were produced, the text refers to seed bearing fruit being produced by these trees. Any horticulturist knows that fruit-bearing trees require several years to mature before they produce fruit. Note the text states that the *land* produced these trees (indicating a natural process) and that it all occurred on the third day. Obviously, such a "day" could not have been only 24 hours long.

6. Is it really reasonable to believe that Adam named the animals, had time to get lonely and meet Eve all in one day? Being sinless does not make one superhuman. Even if Adam only named the animals in Eden, there still would have been quite a few. From Genesis 2:20 it seems apparent Adam was in need of a companion. With God and all those animals, would Adam really get lonely in only one day? In the overview of the creation of humans in Genesis 1:26–29, and the details in Genesis 2, God gave a lot of mandates to Adam and Eve. Also consider that Adam was not created in Eden (Genesis 2:7), so did time pass before he was placed in Eden (in Genesis 2:8)? In any case, it seems more reasonable for days, if not years, to have passed from these verses to the events recorded in Genesis 3.

7. Eve's childbirth pains were *increased* (Genesis 3:16) after the fall of man. This tells us two things: 1. There was pain before the fall (Appendix D will address this more); and 2. She may very well have had kids before Cain and Able. Various clues in the text indicate that the Cain-Able-Seth births may not have been sequential or necessarily close in time. If Cain is able to find a wife so "soon" in Genesis 4:17 and able to build a "city," then other siblings must have been born close to his age. Genesis 5:4 indicates that Adam and Eve had many kids not listed. The timescale in these passages is obviously compressed, the events do not make sense otherwise. One could also conjecture that births might have occurred before the fall, which would also make the growing population afterwards easier to explain.

8. Are God's days the same as our days? No, because as the creator he would be outside of time by virtue of the fact that he created time along with the rest of the universe's attributes. The Bible alludes to this in Psalm 90:4 and 2 Peter 3:8 which seems to foreshadow future discoveries concerning the universe's multi-dimensional structure (see Appendix A).

9. We see that when God rested, he ceased creating and each "day" previous to that was closed out. The seventh day is *not* closed out like the others. As each of the previous days represent eras before man (and the sixth includes early man), the "seventh day" is mankind's entire existence up to and including the present. The Bible speaks of the Sabbath not being closed out (as indicated in Hebrews 4) until the new creation when God starts creating again (Revelation 21).

10. Genesis 2:4 reads, "This is the account of the heavens and the earth when they were created. When the Lord God made the earth and the heavens…" This is the NIV translation, but in this case the KJV better renders the verse in accordance with the Hebrew. The Hebrew for this verse contains yom (day) which is left out of the NIV. Read the KJV: "These are the generations of the heavens and of the earth when they were created, in the day that the Lord God made the earth and the heavens…" This verse serves as a summation of Genesis 1 and an introduction to Genesis 2. If the creation days were 24-hour days, why would this verse refer to the creation as "in the day" God made? This is a clear usage of the "long period of time" definition for day. A similar usage is found in 2 Peter 3:10–12.

11. The Bible makes clear statements on Earth's antiquity in 2 Peter 3:5 and Habakkuk 3:6. Allusions to an ancient world can be found in Psalm 90:2–6, Proverbs 8:22–32, Ecclesiastes 1:3–11 and Micah 6:2. These verses are also interesting because they seem to confirm what many people perceive. Many (most?) people look around them at the natural processes, geologic formations and so forth and find it hard to envision the world being very young. The signs of age are all around us.

12. Exodus 20:11 is often held up as undeniable proof of 24-hour creation days. If that is true, what of Leviticus 25:1–4, which uses the creation week pattern in terms of *years*? Apparently the creation week is used as a pattern of "one out of seven" in both cases, not a real-time reference. A similar type of pattern is the eight day "Feast of the Tabernacles" in Leviticus 23:33–36. It celebrated God's protection in the desert that lasted *forty* years—obviously eight days is not a one-to-one correlation with forty years.

13. Both 1 Chronicles 16:15 and Psalms 105:8 refer to God commanding his word to "a thousand generations." This seems to confirm that the genealogical listings were incomplete. A "thousand generations" also seems to roughly confirm scientific datings on human origins (see Appendix C).

The Point

These very simple points indicate that the creation week was not made up of 24-hour days. We considered simple reasoning and context, but not so simple that we were reading superficially. There are many more issues with the young-earth creationist view, but it is beyond our scope to address every one of them. Appendix D discusses some additional creation-date issues (Death Before Adam, Dinosaurs, Appearance of Age, Why an old Universe? and The One-Sided Equation).

The point here is to show that young-earthism is not based on sound science or biblical scholarship, but on preconceived beliefs fit onto science and the Bible. Based on the fallacy that old age equals evolution and that 24-hour days are *the* literal interpretation, creationists have claimed Earth is young (with no evidence) and that the Bible supports this (with very problematic interpretation). This is why their critiques of Neo-Darwinism have been minimized when naturalists and skeptics point to the problems of "young-earth evidences" to question their credibility. It also directly feeds the belief promoted by skeptics that young-earthism is proof that Christianity is not based in reason and fact and *this* is the foremost reason why young-earthism needs to be addressed.

This problem can be reworded as the question, "How can I believe the Bible or in Christianity when it claims Earth is only a few thousand years old and was covered by a global flood?" This is a major stumbling block for skeptics *and* Christians, which is why Christian scholars should focus on solving this problem. It is not uncommon to hear accounts of people who struggle in the separation of their "religious life" from their career or the science they are exposed to because of the contradictions young-earthism causes[6] (such as the claim that Earth is only 6000 or so years old when civilization is *easily* traced further back than that). Instead of resolving this issue, too many blindly choose sides or sweep the problem under the rug (as we will see the intelligent design movement tends to do).

Some young-earth creationists have become so vocal and persistent in their views, they are often seen as being representative of Christianity. Emotional or "just-so" responses seem to be common from the young-earth side. For example, a recent article published by young-earth group Answers in Genesis declares the big bang wrong.[7] It does not answer the evidences for the big bang, but simply says it is wrong because their *interpretation* of the Bible necessitates that the big bang must be wrong. This does not refute the big bang and it also begs the question of what support exists for their young-earth interpretation. The article also declares Christian supporters of the big bang "compromisers."

Is labeling people, or name-calling as it may be considered, scholarly? Is it the biblical way? No on both counts. Instead of discussing the issues, we find articles like "Warning to Families![8]" which does nothing to address the issue, but is designed to raise emotion over "bloodshed and killing" before Adam (see Appendix D). Or in "What's Wrong with 'Progressive Creation'?[9]" (progressive creation refers to old-earth creationism) we find statements such as "tear-eyed mother of young children" and "I don't know who to trust anymore" rather than sound, scholarly discussion. Reviewing young-earth literature one could document many more examples such as these.

There is a definite pattern of emotion used over rational discussion, but not all young-earth materials are like this. Author, president of the Christian Research Institute and young-earth supporter Hank Hanegraaff often discusses that salvation is not contingent on one's stand on the creation-date issue. He often promotes reasonable discussion on the creation-date issue on his *Bible Answer Man* radio program, but when he aired programs with a well-known old-earth creationist, he was "deluged with acrid criticism[10]" from young-earthers. Young-earth creationist Dr. John Mark Reynolds summed up this problem by saying, "I think too often we adopt a sort of fragile or brittle approach to this sort of thing where we end up yelling at each other and being divisive inside the body of Christ. Only the naturalists benefit from that.[11]"

Young-earthism is also prevalent in many Christian-based home-schooling curriculums. Home schooling has shown it can be an effective education method. However, many home-schoolers do not undertake the necessary study and research to understand science fully and blindly trust curriculums with a "Christian" label. It is not the label that counts, but the quality of scholarship that ultimately determines a source's worth. Teaching science through the filter of young-earthism will cause as many problems as curriculums that teach naturalism. Both do not teach science, but teach what certain people want science to say. Too often people rely on what they were always *told* was right, not what they found to be correct after research and study.

It is in young-earth books that a particular contradictory tenet of young-earthism becomes apparent and sometimes confusing to the readers. While on one hand teaching that most of fundamental science (dating methods, the big bang, and other things we have talked about) is wrong, science is trusted enough to "prove" the young-earth view. We have an antiscience, "do not trust science" attitude at the same time we are told science is good enough to prove another view and verify the Bible. The young-earth view is also unable to use the strong evidences the old-earth view does concerning intelligent design theory and Bible-science compatibility discussions (such as the big bang, fine-tuning of the speed of light and Genesis-science agreement). One can easily find people who claim old-earth science is all wrong, yet do not know why other than from "pamphlet theology" or from what people have told them is correct. It also seems that a large percentage of young-earth supporters have little or no scientific background. This is not a slight against them, for it can be said of most people, but how does one make claims about the validity of one theory over another when the very basics of *undisputed* science are unknown to them? More importantly, how can one make a claim about *anything* in good conscious when they have not studied the subject or have based their belief entirely (or nearly so) on what someone told them was correct? We need to abandon such childish thinking and replace it with mature, adult minds.

Naturalism, at least, does not readily abandon fundamental, well-proven science. Its problem lies more with methodology—the way it practices and interprets science. Young-earth creationism goes beyond methodological problems and attempts to recreate science entirely. It then force-fits this new science onto the Bible, with the results we have discussed. Staying true to its roots, Christianity should not ignore such issues. It often focuses on cults, general theology and nonscientific apologetics while largely ignoring science issues.

This is not to say that these creationists have not contributed productively to the debate on origins. Not all are driven by emotion and uncritical thought. However, the importance of this issue and why it should be addressed was summed up best by David G. Hagopian:

> This debate has important ramifications for how we interpret Scripture, proclaim the faith, embrace science, and stand on the shoulders of those who have preceded us in the faith. We all would do well to remember that we agree on far more than we disagree, but we also must remember that we gain nothing by ignoring our differences or sweeping them under the rug. In fact, we stand to gain quite a bit by discussing our differences openly, honestly, and charitably.[12]

Test what you believe, do not blindly subscribe to whatever you are told. Let reason and science lead to where they do and truth will be found.

15

A Designed Universe?

We have discussed why it is illogical and unscientific to tell science what it may or may not discover. We have seen the flaws that serve as a foundation to philosophical naturalism and naturalistic-influenced science. Christian theism's compatibility with science, and ability to address scientific issues, has been found to be secure once we strip away misconceptions and erroneous creation science. Where does intelligent design theory fit into this?

Already, we have detailed evidences intelligent design theorists use such as complex specified information, irreducible complexity and fine-tuned parameters in the universe. Such things are unexplainable by chance driven processes. As strong as these evidences are, is design as easy to detect as supporters claim or is it harder as skeptics suggest? Can the intelligent design theory be formulated as model that makes predictions that can be tested or is it untestable speculation or based on philosophical contradictions like naturalism?

Detecting Design

The discussion on complex specified information (CSI) in Chapter 13 described how order can exist in nature and whether or not chance can produce information. We concluded nature could never produce complex specified information, only complex unspecified or noncomplex specified information (which are technically not information, but rather patterns that superficially seem like information).

The irreducible complexity of biochemical systems differs greatly from the general order seen in nature. A snowflake takes on an ordered appearance. That order itself is a result of natural laws and contains no information. On the other hand, if the laws and forces that produce that snowflake were deconstructed, one would find the same precisely fine-tuned laws that govern life's existence. Any particular biochemical system runs into this wall of complexity far sooner and is much easier to detect. Consider the analogous spacecraft. It is ordered and assembled in such a way that nature could never produce it, even if its parts already existed "as is" in nature. The spacecraft's specified order and complexity point to intelligence. Biological systems, as we saw in Chapter 13, also have this level of complexity.

Another thing to consider is *contingency*. Contingency means "dependence." If an object, event or structure is considered contingent, that means they are *compatible* with underlying natural laws, but *not required* by them (the object, event or structure does not unavoidably have to happen because of those laws).

This may be headache inducing, but think about it. If you ran across a message in the sand, you would immediately recognize it as being *caused* by an intelligence. A cloud that looks like an animal, on the other hand, you relegate to unintelligent wind. The former example is a contingent, complex, specified event. The latter is an uncontingent, necessary, unspecified, complex event. It is not merely compatible with natural laws, it is *required* by them. One is caused by intelligence, one is not.

The ability to detect intelligence is common to all people. So common in fact, that we use it every day. Whole fields of study are based on it such as forensics, archaeology, cryptography and so forth. Efforts to discover extraterrestrial life (known as SETI: the Search for Extraterrestrial Intelligence) rest on the ability to detect design. Ironically, SETI efforts are driven by naturalists looking for the vindication of their worldview and Neo-Darwinism that a life-filled universe would provide. Detecting design is not some highly complex or miraculous process, it is a simple and very common process inherent to the human race. Mathematician William Dembski describes this ability this way:

> Intelligent design is a theory for making sense of intelligent causes. As such, intelligent design formalizes and makes precise something we do all the time. All of us are all the time engaged in a form of rational activity that, without being tendentious, can be described as inferring design. Inferring design is a common and well-accepted human activity…There is no magic, vitalism, no appeal to occult forces. Inferring design is common, rational and objectifiable.[1]

Intelligent design theorists have been driving the point home that detecting design is a scientific process. If that process ultimately points to a designer, then that is what it points to. Most have stopped at that point, saying little about the nature or character of the designer. They have by and large left this to the theological arena in order to avoid the "disguised creationism" label given to them by naturalists. They also want to avoid the historic "government funded schools cannot teach religion" spectacles. Young-earthism is rarely seen in public schools because it is more theology than science. On the other hand, naturalistic theology manages to disguise itself as science all the time. The difference with intelligent design theory is that it centers on practicing good science first, implications later. Do those implications (who is this designer?) have to wait?

No, because if you follow the science to where it leads, eventually you have to discuss who this designer is. That is the next logical step. You cannot simply say there is a designer and leave it that. Indeed, this is where science and religion converge, but it is still science. For intelligent design supporters to stop short of the implications of their theory is to turn their back on the good science practices they are so intent on promoting. Their attempt to avoid conflict may be reasonable at first, but it will only cause more problems later on. A vague designer or being is only inviting relativistic theories and beliefs to define this designer however they deem necessary.

The Critics Respond

Nevertheless, intelligent design has naturalists worried. Their once exclusive focus on young-earthism now often turns to design supporters. Their comments often concede to the sophistication of design and the credentials of the people behind it. The media pays attention to intelligent design and less and less considers it to be some sort of religious movement. Design critic Robert Wright writes, "Instead of being a bunch of yahoos, they are a bunch of 'academics and intellectuals' with new, 'more sophisticated' ideas.[2]" The naturalists know what they are up against and have turned to portraying design as "disguised creation." Since they consider creationists "yahoos"—and have spent years painting them as such—what better way to minimize intelligent design than to convince people it is the same old thing? The problem with this approach is that most people see it for what it is: A mischaracterization, stereotype or straw man argument that does not address any of intelligent design's arguments. This restating of the design arguments in such a way to make them easy to deconstruct is an obvious fallacy. Yet it is the common method of attack. We will continue to use Robert Wright's article as a case study, since it seems representative of design critics.[3-4]

He claims Phillip Johnson, author of *Darwin on Trial*, is "suffering from an elementary confusion about Darwinian theory." Reading carefully, the author does not refute anything, but talks around the point Johnson actually made. That point being, while natural selection does occur at some levels, it does not serve as the miraculous mechanism to produce common descent or new animal types from another (say a mammal from a reptile). Relatively minor adaptations are one thing, completely new biochemically complex organisms are another.

The author then tries to minimize Michael Behe, author of *Darwin's Black Box*, by simply pointing out that Behe's arguments are nothing new. This seems rather silly because the points present serious problems for Darwinism, regardless of whether or not they are as old as Darwinism itself. Behe simply applies modern biochemistry to the issue, making the problem worse for Darwinism. That problem is the question of how could chance-based Darwinism ever hope to produce complex, interconnected biochemical systems such as blood-clotting or the eye through hit and miss processes? The question remains unanswered. Both of Wright's simple and technical sounding answers never actually answer this question, but talk around it, perhaps hoping people will not notice the fundamental question remains unanswered.

Wright then addresses William Dembski, a mathematician with a number of books on intelligent design. Once again the author speaks around the problems Dembski poses for Darwinism. Dembski has created a process or "filter" to detect design. Wright oversimplifies this process, saying it only detects complexity and that natural selection does not deny complexity. The problem with this is that Dembski takes the process much further: It detects *designed* complexity, not *undesigned* complexity. We have already discussed the difference and how they can be differentiated from each other. Once again, the author has missed the major parts of the arguments against his own theory.

Wright makes the interesting statement, "So far as I can tell, Dembski's argument is just an example of something demonstrated time and again in various disciplines at various accredited universities: If you phrase your argument in mathematical symbolism and technical terms, some people, including other academics, can be counted on to lose track of what the exact connection is between the symbolism and the reality it's supposed to represent. Then they may conclude that your mathematical model proves something…" Oddly, he has explained exactly what naturalists and creationists often do. People fall for their technical or scientific sounding arguments because they *sound* that way. Instead of testing what they hear or read, they accept it based on how it sounds or on the perceived authority of the author.

We detailed that detecting intelligent design is not a mystical process and thus this design cannot be labeled as "apparent design." Another common objection is trying to compare intelligent design with optimal design. Critics may ask, "If God designed it, why isn't it perfect or optimal?" Theistic evolutionists may add,

"Why would God interfere in natural processes that he set off? Did he make mistakes that needed fixed?" These questions assume at least two things: 1. We know the mind of God and what he considers to be optimal; and 2. We assume optimal is defined as perfect or ideal. Both of these assumptions are extremely flawed.

First, assuming to know the mind of God seems to be irrational unless one considers themselves an omnipotent being not bound by the universe's properties. Secondly, any engineer will tell you that optimal design would be better described as *constrained* design. All designs must find the best compromise to meet goals, regulations and costs. Only unlimited money would allow someone to approach perfect design. In nature, an optimal design would best meet that organism's needs once all surrounding influences are taken into consideration. However, what seems to us to be a compromise may only be so from our limited perspective. The discussions on complexity show how interconnected and vast those considerations are for life to exist. So what we see as imperfect or a compromise would most likely be optimal, perfect or ideal if we were able to view the world from a larger perspective.

Why would God interfere with nature and create new life suddenly and fully formed as seen in the fossil record? Is this simply a "God of the gaps" argument? Both theistic and nontheistic evolutionists would say that intelligent design is a gap argument. Then they turn around and talk about the "hope" they have for evolution evidences and discoveries that the future *may* provide. We have seen the impossibility of chance producing intelligence, the ease of detecting designed complexity and so forth. Which is more scientific and empirical? Hope or fact-based theories and models? The critics' "God of the gaps" claims are simply camouflage of their own "appeal to ignorance" fallacies.

So the whole issue once again boils down to testing what you read, detecting the fallacies and digging deeper beyond what the superficial media reports detail. Even if many of these reactions to intelligent design are flawed, and design itself is easily detected, can it truly be formulated as a scientific theory? Any good theory makes predictions that can be tested and verified.

Intelligent Design as a Testable Scientific Theory and Model

We briefly mentioned William Dembski's "filter" that detects whether something is intelligently designed or is produced by chance. This filter is an important part of intelligent design theory. Remember our discussions on contingency, specification and complexity? Dembski formulated these into a filter to apply to objects, events or structures.[5]

1. Is it contingent? If No, then it is produced by necessity. If Yes, go to 2.

2. Is it complex? If No, then it is produced by chance. If Yes, go to 3.

3. Is it specified? If No, then it is produced by chance. If Yes, go to 4.

4. It is designed.

This is one way to formulate our every day design detecting abilities in a more scientific manner. Any good scientific theory goes further and makes predictions of what will be discovered later or should be found now. The intelligent design supporters have been noticeably lax on this important part. This seems largely due to the fact that once you create such a model, the "whys" and "who" behind design lead to the religious implications some design theorists try to avoid. Astronomer Hugh Ross has attempted to formulate a "creation model" without worrying about the resultant theological implications. Instead he embraces those implications as part of the model, because that is where his model and intelligent design ultimately lead. This model says the following should be found (or continue to be found or supported) in nature if the universe was designed.[6] There are more points in the model, but these are for the most part ones we have discussed in some form in this and previous chapters.

1. Fine-tuned transcendent creation event where all matter, energy, space-time begins (big bang).

2. Cosmic fine-tuning.

3. Fine-tuning of Earth's, the Solar System's and Milky Way galaxy's characteristics.

4. Rapidity of life's origin.

5. Extreme biomolecular complexity.

6. Cambrian explosion (sudden appearance of most life types during same time period).

7. Missing horizontal branches in the fossil record.

8. Lack of "transitional forms" in the fossil record.

9. Frequent mass extinctions (to allow new life forms to be introduced).

10. Rapid recovery from mass extinctions.

11. Recent origin of humanity (as opposed to common descent).

12. Huge biodeposits (needed to sustain humanity).

13. Molecular clock rates which show humanity's relatively recent origin.

Intelligent design is not some new-fangled theory. Many people over the centuries, perhaps even you, have looked around at the complexity and variation of nature and concluded that chance could not produce such things. People often look around them and ponder thoughts like, "How could chance produce such complexity? Dinosaurs didn't just one day decide to become birds and 'poof' there they were. How could chance produce the diversity we see from one ancestor? How does chance produce so many convergent traits in animals such as birds and bats that are at the same time very different biochemically? How can chance produce some animals that are virtually identical, yet genetically unrelated?" And so on.

Intelligent design seems to be the logical answer and now science supports that conclusion. These points not only support the idea that intelligent design is inherent in the universe, but that humans are the end reason of *why* the universe was designed. Sooner or later this "Who designed and why?" comes into play. Design supporters want to avoid this for the time being in order to establish design as a science in face of the "religion and science are separate realms" paradigm. Instead, we should let the science guide us to whatever destination it leads.

One place where the science leads is to irreducible complexity. We referred to this in discussing the complex, interconnected systems in the giraffe and the impossible changes required to transform one organism into a completely new one. No where is this complexity more evident than in the cell.

The complexity of cells becomes apparent under extreme magnification which reveals their structure. Take, for example, the bacterial flagellum. It has parts referred to as the propeller (or filament), rotor, drive shaft (or rod), bushing, universal joint (or hook), etc. These are obviously names from mechanical devices, but they are not used simply because they are convenient analogies. These components are precise biological versions of their human-designed mechanical versions. In fact they are more efficient and precise than anything we could design. Nor could these cells be simply formed from existing "parts" from other cells.

Each cell has a unique structure, precisely intended for particular functions, even those that have a few parts common to other cell types. In other words, if you were able to enlarge one of these cells and leave it lying in the woods, someone who found it would recognize it as a designed object.

It is just such an object that Charles Darwin said would undermine his theory. In *Origin of Species* he wrote, "If it could be demonstrated that any complex organ existed which could not possibly have been formed by numerous, successive, slight modifications, my theory would break down.[7]"

Intelligent Design and Christianity

We have already looked at how Christian theism is rooted in reason and is not antiscience, so where does intelligent design come in to play? Many of the leading design theorists are Christians who believe the designer is God of Christianity. They do not hide their beliefs, but as stated before they downplay the religious side as a way to establish design as a science in the face of the hostile "science versus religion" mindset of many.

Perhaps a better approach would be to make religious beliefs perfectly clear, so critics cannot claim "disguised creationism" or hidden bias of some sort. Remove the basis for these hostile attacks, then say, "Here's my scientific theory. Don't claim it's invalid because you disagree with my beliefs. Put it to the test scientifically instead of throwing illogic at it." Then demand all types of naturalists and creationists abide by the same level of scholarship.

Christian design theorists should also stop avoiding related issues such as the young-earth paradigm. Young-earth supporters of intelligent design find themselves between a rock and a hard place. Most leading intelligent design theorists find validity with the evidence that the universe is very old. Their books do not question the age of the universe. The science that proves the universe is ancient is the same science that is the basis for intelligent design theory. So while young-earthers embrace some of intelligent design, they are highly critical of other parts and certain methodology simply because of the old age component.[8] In the process they have to abandon the strongest parts of intelligent design including the big bang (some design theorists will even avoid the big bang in order to avoid angering young-earth allies).

Intelligent design could address the creation-date issue in order to remove a stumbling block in the origins debate. They would no longer have to avoid evidences that are related to old age. No longer could skeptics point to young-earth beliefs of some design supporters as a reason to call intelligent design "disguised creationism." Nor could they point to creationist disagreements as a reason not to trust their design theories. The underlying problem is that design theorists have embraced a "big tent" philosophy that accepts all sorts of design theory backgrounds. In effect, this sometimes makes the theory harder to sell and invites future problems. Science should not be driven by such politics. Anyone truly interested in finding the truth will not get offended.

Not addressing internal issues will lead to relativistic thinking. Even if intelligent design is as solid as it seems, some may try to use it to "prove" the existence of any god or designer that they see fit. True science will seek to avoid illogical relativism by not avoiding the hard questions. One such question is, "What religion can point to intelligent design and say, 'This points to our God'?" or "What is the nature of this designer?" In other words, if science points to a designer, the next logical step of science would be to attempt a determination of the nature and attributes of that designer.

Many believe that answer would be Christian theism. Material in Appendix A and B details some of the Bible's scientific accuracy, written long before the relevant discoveries in science. Even if the necessity of a designer is apparent to you from the evidence, maybe some questions like "Where did God come from?" still trouble you. Appendix E attempts to address that age-old concern.

Other religious books are written in human terms based on the knowledge of the time. Creation stories concern themselves with creators bound by the universe or reveal inconsistent knowledge about nature. The Bible seems written with information unknown at the time (see Appendix A for a discussion on the trinity, which serves as a perfect example). Chance? Coincidence? Or is one religion really based in factual reality? Modern science and critical study can point to inconsistencies in religions. What about Judaism and Islam which share similar roots to Christianity? Begin where they diverge from Christianity and test them. Which are consistent? Which contradict? You get the picture.

Going through all that testing here would be impossible. The Notes for Chapter 9 list resources that address both religious and scientific issues relating to Christianity and other religions. However, by only looking at the science side here, Christianity seems to have an "unnatural," or *supernatural*, agreement with science. If God of the Bible (assuming for a moment that all your questions about the Bible have been adequately addressed) did create the universe and inspire the Bible, then logic suggests science should reveal evidence that agrees with biblical writings. Intelligent design is one such evidence. That evidence specifically reveals a creation specifically tailored for humanity's existence.

A true worldview will not contradict itself. A true worldview will fit the facts. One can live consistently with a worldview that is true. Naturalistic and relativistic-based worldviews fail this. The science they produce fails testing. Christian theism passes these tests. It does not contradict science. Consequently, it is not compatible with naturalism and relativism as some try to reason. There is much more to be studied to these ends, but hopefully a road to those ends can be seen.

16

Connections

"All we have to decide is what to do with the time given to us."—Gandalf, *The Lord of the Rings: The Fellowship of the Ring* [1]

We have completed our whirlwind tour of critical thinking and ferreting out pseudoscience. Our focus was on commonly encountered issues, mainly science related topics. Any one of these chapters could be expanded into volumes detailing their subjects, but it was not the goal of this book to comprehensively address each topic. The line had to be drawn at some point to prevent this book from being two thousand pages, which no one would neither buy nor read. This book attempted to cut through much of the haze and zero in on the substance and main points of each. So at one level this book is necessarily incomplete. If any particular subject is of interest to you, study it further, test it and try to understand it.

Another theme of this book is *connections*. Virtually everything, if not everything, we study, hear and read are connected in some way. Did you make the connections between the chapters in this book? At the very beginning, the maxim "Truth is Not Relative to the Beholder" was presented as the major connecting theme. Some connections are subtler, others are more obvious.

Consider the chapters themselves. Chapter 3 establishes the importance of factual history, which leads into the next chapter's study of mankind's inherent corrupted nature. This nature preys on the gullibility of people (Chapter 5) which in turn is based in part on a poor scientific background (Chapters 6–8). That takes us to the relations between religion, science and reason (Chapters 9–11) which in turn serves as the background for the questions concerning the universe's origins in Chapters 12–15.

Other connections may be less subtle. Chapter 4 detailed the direct connections between historical events. The discussions on origins drew on many scientific disciplines from physics to biology to archaeology and anthropology and then it converged with theology.

We saw how one practices science is often influenced by the theologies and philosophies of that person. If you believe man is inherently good and corrupted by society (Chapter 4), naturalism is your best bet because "survival of the fittest" provides justification for evils. Naturalism can give you justification for radical environmentalism (Chapter 8) by lowering the importance of human life (because naturalism says we are formed by the same "chance based forces" as all other life). Yet we also saw how these various justifications from naturalism are circular and contradictory. Such rationalizations are often not based on facts, logic and objective truth.

Hopefully you can see how everything in life is connected. Issues do not live in isolation from each other. You may now see or hear a subject discussed and ask, "What is his reasoning? What are his influences? What proof exists for his conclusion?" Logical fallacies in articles or from politicians may stand out more. The importance of always educating yourself may also be apparent. Connections between one subject and another may no longer be hidden.

Many scoff at the idea of studying anything other than what relates directly to their job or career. Some will not even do that. "Why should I study history, science, economics or politics?" Would you not like to be able to explain such things to your kids? Answer their questions correctly? Make correct decisions? Vote for people that best serve their constituents? Learn more about who you are and why you are here? Instead of waking everyday and going through the grind blindly, is not better to learn what purpose there is to life and contribute to it and to those who will follow in the future?

Maybe this book will encourage people to determine why they believe the way they do. Perhaps the subjects herein have raised an interest in a particular topic to be followed by more study. If you do not agree with everything in this book, instead of shouting "No, you're wrong!" and waving your hands or feeling offended, maybe you will step back and say, "Why do I believe what I do and can I prove it?" There is nothing wrong with disagreement, it is the manner in which that disagreement is resolved that reveals one's ability to think clearly and function as an adult.

If nothing else, I hope readers will approach all things knowing that a little more effort and a little more critical thought can bring greater understanding of the world. Step back from the busyness of life and determine why and where you are going. What worldview do you live by? Is it consistent? Logical? Defensible? If not, what are you going to do about it?

Time can be seen as a gift. Time, however, is not unlimited. Use it wisely for the betterment of the world. Critical thought, reason and education can help you rise above everyday life and leave something behind. That "something" may be what you produce or accomplish. Perhaps it will be how you live or how you raise your children.

As Maximus said in the movie *Gladiator*, "What we do in life, echoes in eternity." This notion is rooted in reason and is often repeated in both secular and religious society. It can be taken to refer to what you leave behind in this world, what you will face in the next or both. However you take it—and regardless of your religious beliefs—the certain thing is that you cannot ignore it. To say that your life and actions mean nothing is to speak irrationally. One way or another, your legacy will either haunt or benefit those you leave behind. Maybe it will haunt or benefit *your very being*. Reason underlies reality. Something then must logically underlie reality because reason suggests intelligence. So the "something" must be a "someone."

Reason is only the beginning, not the end.

Appendix A

Beyond Reality

In Chapter 5 it was mentioned that most UFOs can be reasonably explained away. A small percentage are left over as unexplained, or "residual UFOs" (RUFOs). These seem to happen to people involved with occult practices. Throughout history, similar RUFO accounts can be found, but they are in terms of that particular culture and time. Today they are aliens, then they were gods, trolls, fairies, demons or spirits.

People who research the paranormal or supernatural find "ghost stories" or "demon encounters" throughout the ages. Often the ghosts are seen or felt where something tragic occurred. Demons dwell in places associated with occult symbols, rituals or activities. When all is said and done, not all such "paranormal events" can be explained away, hence something that can be classified as "residual" as with the RUFOs.

This becomes another indicator for a "hidden" part of reality. In Chapter 15, we looked at scientific evidence for a designer behind the universe. All of this only raises the question: Can science be a little more specific in regards to a "nonphysical" part of reality? Sure, intelligent design theory may be compelling, but what else does science say?

Multiple dimensions are often the realms of science fiction. Modern physics brought these dimensions out of fiction when it found they are integral to the fabric of the universe. We are familiar with our everyday three-dimensional world. Most are also aware that physics considers time an additional dimension, but one that can only be traveled along in one direction. If you mark the height, width and length (the spatial dimensions) of a cube, you notice the three lines are perpendicular to each other. Time (the temporal dimension) would be perpendicular to each of those three. Together these four dimensions are known as spacetime. This is impossible to visualize other than through mathematical terms. However, Einstein, through his relativity equations, discovered time is integral to the fabric of space and must exist as a dimension to explain how the universe functions. Also recall our discussion on how mass "warps" the *space* around it. Both time and space in a sense have substance, but not in the sense we are used to. That is nonphysical reality.

> **Consider This**: Before science realized time was a dimension, H.G. Wells described time as such in his science fiction book *The Time Machine*. The book's time traveler essentially asked his friends if they agreed that any real object must have height, width and length. They agreed. Then he asked them if such an object did *not* exist for any period of time—it was instanta-neous—would it still be real? No it would not be real because a real object is defined by time—or *duration* as Wells called it—as much as it is by the three spatial dimensions.

Scientists for decades have been trying to reconcile different sets of physical laws. Quantum mechanics successfully describes the minute, relativity describes the vast. They do not do so well if the arena is reversed. Throw the forces of grav-ity, electromagnetism and nuclear forces into the mix. They should at some point all converge, agree and form part of one theory. Called the "Theory of Every-thing," this is the Holy Grail sought by physics. As our technology and knowl-edge increased, scientists found something interesting to these ends.

For all of these well-tested scientific laws and forces to be reconciled, more dimensions had to be introduced. For the universe to exist as we find it today, at least six additional dimensions have to exist beyond our own four. Those dimen-sions exist "wrapped" around our own: Not here or there but everywhere. Can beings exist in or beyond those dimensions? Probably not in those dimensions themselves because of their nature (minute and attached to our own), but their nature is not fully understood. The point is that we can be certain that more exists in this universe than meets the eye and we do not fully understand the nature of the reality we cannot see.

Dimensions and the Creator

Logic suggests that what-or whoever set the universe off would have to be sepa-rate from the universe. This is similar to how a mother is separated from her baby at birth or a craftsman is separate or "outside" of his creation. So even if a being did not start off the universe, there must be existence beyond our own ten or more dimensions. We started off from something, but at the same time we must exist independent of the cause. To put this into perspective, consider the follow-ing illustrations.

Look down on a globe, hovering above the North Pole. You are some being beyond our universe and its many dimensions. Pretend the equator has a stopping and starting point and consider this line the dimension of time. Follow longitude lines down to the beginning, middle and end of the line you created on the equator. These points represent events in someone's life: Birth, life and death. You can see them all at once or interact with them one at a time. Outside of the universe, you are not bound by the dimension of time or any other dimension. If this is true, you can interact with any of these dimensions.

Consider that normally you can perceive and interact with something that is two-dimensional such as a two-dimensional drawing. You cannot interact with something in the sixth spatial dimension, only something within our own dimensions. Any "something" in a dimension outside our own has to enter our own dimensions before we can interact with it.

While we have left out the technical and mathematical parts of multi-dimensional physics, we can conclude a couple things about our universe. First, the structure of the universe leaves room for what we perceive as "supernatural." Second, this physics also explains to some extent how a creator can be omnipotent (being unbound by time, as in our globe example). Science no longer can be used to toss out the supernatural. What science can and should be used for is attempting to determine which of these events are supernatural and which are not.

There is another interesting explanation physics presents for the nature of God. Specifically, in Christianity, God is described as existing simultaneously as three distinct persons (God, Christ and Holy Spirit) yet at the same time is independent from the two other parts. This "one = three" or "three = one" is seen as an impossibility by critics, including Jehovah's Witnesses and those who practice Islam. In our everyday world, one cannot equal three, but as we have discussed, there is more to our world than meets the eye.

Hold up a piece of paper and pretend it represents our three-dimensional world. Cut a hole in the paper and pass one finger through it. This represents an intelligent being from beyond our existence. Cut another hole and place a *different* finger from the *same* hand through it. These fingers look and act like *different* beings from our perspective (albeit similar in many ways), yet they are part of the same being from a larger perspective. From our point of view, two is not equal to one, but from the larger perspective that includes other dimensions, the contradiction can be resolved.

One last point, recall the Psalm 90:4 and 2 Peter 3:8 indicate that God's time is not the same as ours. These verses can seem confusing or contradictory. Now that science has shown there is existence outside of our time, these verses make sense for a being that is not bound by our time (more on time and God in Appendix E).

Questions

While some want to find a demon behind every bush or a ghost behind every noise, the preponderance of evidence indicates there is more to this world than only people. Even if you throw out every slightly questionable event, you are still left with many occurrences. "Haunted" places seem to be found where traumatic events occurred (such as murders or battles) or where people have lived for many centuries. Some people seem to be more "in-tune" to such phenomenon, just like some people have a better immune system than others. When it comes to "evil spirits" or demons, a lot of these events seem to occur when the people invite such activity through their occult practices.

Do certain events or activities create particular connections to other dimensions or existences beyond our own? Do some unknown parts of physics explain how a person can leave behind a "presence?" Does a sudden release of energy through traumatic death disturb the fabric of space and time? Have malevolent beings from another existence been terrorizing people for centuries? Countless people have had experiences that seem to indicate that there is something tangible to these questions. History seems full of "otherworldly" mythical events or "mystical" places which seem very similar to each other, with each time period describing them somewhat differently on the surface in the terms of the day. Physics shows that the universe does exist and was created in such a way to allow these "nonphysical" realities. We can no longer sweep this fascinating, and sometimes frightening, line of thinking under the rug.

It is another place where science and religion converge. This is also an area where reason and critical thought can be helpful in determining which realities are in fact real and which are fakery or misunderstanding.

APPENDIX B

Science in the Bible

Our discussions on science vs. religion (specifically Christianity) and figuring out which religion's creator that intelligent design is referring to are contingent on whether or not the Bible agrees with science. We discussed some resolutions to the problems that the global flood and young-earth theories produce. In that process we came across some interesting biblical scientific claims that predate the actual discoveries by thousands of years.

Psalm 104 reveals that early Earth was completely covered by water (as does Genesis 1) long before science determined Earth had such a period. The apparent contradiction of light appearing before the Sun's creation in Genesis turned out to be a contradiction only in the young-earth view. In Chapter 14 we saw how a reasonable resolution to this contradiction revealed the Bible spoke of a time in Earth's history long before scientists came to the same conclusion. That same chapter also listed biblical indications of Earth's antiquity. Another interesting "coincidence" is that Genesis lists the arrival of each form of life in the same order science has established. Appendix A showed that the "contradictory" trinity that is fundamental to Christianity is not so contradictory and reveals an understanding of science unknown until the Twentieth Century.

Other religious books are written in human terms, while the Bible seems written with knowledge unknown at the time of its writing. The following verses are a collection of such examples that either confirm or predict scientific knowledge unknown to the writers or that contradict beliefs of other peoples alive at the same time. This list is by no means comprehensive, it could be much longer, but this is enough to ask the question: Are the verses coincidence or divine inspiration? Why is there so much convergence of science with the Bible in this time in history? Decide for yourself (The Notes for the various relevant chapters list resources that detail this topic at much greater length).

Creation of the Universe: Genesis 1:1, 2:3, 2:4, Psalm 148:5 and Isaiah 40:26, 42:5, 45:18. Interestingly enough, the Hebrew word used for "created" in these verses means something created that was entirely new. This is contrary to various creation mythologies, etc., that claim the universe always existed. Also, these verses agree with what big bang astronomy confirmed centuries later.

Expansion of the universe: Job 9:8, Psalm 104:2, Isaiah 40:22, 42:5, 44:24, 45:12, 48:13, 51:13, Jeremiah 10:12, 51:15 and Zechariah 12:1. It was not until the Twentieth Century that the universe was found to be expanding.

Earth is not held up by anything and the existence of space: Job 26:7.

Sickness and health: Leviticus chapters 12 through 17 often seem a boring collection of mandates for the Hebrews to follow. Yet they contain numerous references to actions whose purpose seems to be a concern for sanitation and reduction of disease. This was long before the discovery of "germs" and their connection to sanitary conditions. Also consider that the Hebrews had lived with Egyptians for many years and the medical advancement of that culture was quite poor in many areas. For example, medicines that include feces as an ingredient were part of Egyptian medicine. So where did the Hebrews get their knowledge?

Stars innumerable: Genesis 15:5 and Jeremiah 33:22.

Time had a beginning: 2 Timothy 1:9 and Titus 1:2.

Earth is round: Isaiah 40:22.

Disappearance of land bridges at end of last Ice Age: Genesis 10:25.

All people are related: Acts 17:26.

Water Cycle: Job 36:27–28 and Ecclesiastics 1:7.

Variety of Stars: 1 Corinthians 15:41.

Earth was once covered in water: Genesis 1:1–8 and Psalm 104:6.

Formation of land: Genesis 1:9 and Job 38:8–11.

Earth's Antiquity: Psalm 90:1–6, Proverbs 8:22–32, Ecclesiastes 1:3–11, Micah 6:2, Habakkuk 3:6, and 2 Peter 3:5.

Man's Antiquity: 1 Chronicles 16:15 and Psalm 105:8.

APPENDIX C

Prehistory

Here we will establish a timeline of the universe and mankind's history. Some of these dates are constantly in the state of refinement as the threshold of knowledge and technology is pushed back. These refinements, such as the date of the big bang creation event, are relatively small. They are a matter of fine-tuning. Dates for mankind's origin and events in its history have undergone a lot of refinement in recent years as DNA dating techniques come of age and more discoveries in the field are made. The latest research continues to separate humans from other hominid species.

You will find that sources are not always consistent in how they label dates. "Years ago" or B.C. (Before Christ) are most common, but sources will often switch between the two in the same writing. B.P. (Before Present) is also used as a politically correct alternative to B.C. Before Christ is based on an unchanging frame of reference (the year 1. A.D.) and "years ago" or B.P. needs adjusting with the passage of time. The different formats may seem confusing, but are actually only marginally different. Consider this example: You may find some sources listing the end of the last Ice Age as 10,000 years ago or 10,000 B.C., so which is it? 10,000 B.C. is equivalent to a little over 12,000 years ago, which is in the margin of error for the other date. On one hand, the dates are essentially the same, on the other hand it would be nice if everyone stuck with one system, especially one with an unchanging frame of reference.

The margin of error, or error bars, are something to keep in mind with dates of past events. These error bars do not mean that datings are inaccurate, just that distant events cannot be pinpointed to the day or year. However, we can close in on the period of time with some level of confidence. How confident we are is reflected in the error bars. Since most sources usually do not list the error bars, or list ranges that include them, the following list does the same.

13–14 billion years ago: Big bang creation event (also known as the creation singularity).

4.5 billion years ago: Earth formation with Moon formation shortly thereafter.

543 million years ago: Cambrian explosion.

65 million years ago: Yucatán Peninsula impact event. One of five impacts in which more than 70 percent of species died. This is the best known since it killed off the dinosaurs.

2.5–1.5 million years ago: Earliest hominid species capable of using tools.

300,000 years ago: Homo erectus hominids die out.

150,000–100,000 years ago: Most studies have narrowed down human appearance to no longer ago than this date range.

60,000–50,000 years ago: Many genetic studies and recent archaeological data (see two entries below) point to this date as being that of first human appearance.

50,000–40,000 years ago: Sophisticated works of art appear.

50,000–24,000 years ago: Evidence of religious expression found.

47,000–35,000 B.C.: Genetic studies trace the most recent common male ancestor of all humans to this date range. Studies show the female ancestor common to all humans at a few thousand (60,000 entry) years earlier. Chapter 13 discuses these studies in more detail.

35,000–30,000 years ago: Neanderthal hominids cease to exist in Europe and Asia as humans move into these areas (perhaps causing their extinction).

30,000 years ago: First humans possibly reached America. If they did arrive this early, they left little mark and this date is highly debated.

25,000–13,000 years ago: Second Siberian-Alaska land bridge.

14,000–11,000 years ago: Interglacial warming period.

12,000–10,000 years ago: End of last Ice Age and North America continuously settled at least since this date. There is currently a lot of debate on how early man reached North America and what constitutes the earliest settlement date. These dates are most common, but 16,000 years ago or more are suggested by some data. Somewhere in this range seems reasonable.

11,000 B.C.: Domestication of wheat in Northern Mesopotamia.

10,000 B.C.: Land bridges flooded and domestication of "farm" animals.

9000 B.C.: Large-scale agriculture spreads out of Middle East.

Why do genetic studies trace the first human woman to about 60,000–50,000 years ago and the first man to 47,000–35,000 B.C? Perhaps Genesis has the answer. In the account of Noah and the flood, all men on board the ark were blood related, the women were not. So the most recent common ancestor of the men would be Noah. The women could trace their common ancestor to Eve (see Chapter 13 for more on these studies) a few thousand years earlier. Recall that the genealogies in the Bible do not necessarily include every generation (see Chapter 12), so this leaves open the possibility that the flood occurred during these dates indicated by genetic studies. I would speculate that the flood occurred closer to the later date range (47,000–35,000 B.C.), shortly before mankind moved away from Mesopotamia before 30,000 years ago. Recent discussions of a major flood in the Black Sea area around 5500 B.C. became popular through the book *Noah's Flood* by Walter Pitman and William Ryan. This flood seems far too recent and too small in scope and effect on humans to be Noah's flood and to account for the proliferation of accounts detailing a devastating flood.

From mankind's origin to 10,000 B.C. or so is a period of time largely lost to history. This "primeval" or "prehistory" period is when mankind was closer to the wonder of creation (or nature) and possibly the supernatural truths that we so often debate in modern times. Was knowledge greater in those times as some suggest? Were the people guardians of lost technology? Probably not, but a greater understanding of who was behind the creation may have existed. Future discoveries may reveal insight to these times. For now we rely on tidbits of information on this era gleaned from ancient sources that are closer to lost antiquity.

Extra-biblical References

It was mentioned before that many cultures around the globe contain references to a massive flood event that afflicted mankind. Does the Bible contain any other possible references to this past? In Genesis 10:25 there is a brief, enigmatic reference to Peleg's day as the time "earth was divided." Does this refer to the final collapse of the land bridges at the end of the last Ice Age? If it does, we can date Peleg existing approximately 9000 B.C., or 7000 years before Abraham. This gives us some more insight on the gaps in the biblical genealogies (see Chapters 12).

The Bible does quote or reference extra-biblical material, both religious and secular, quite a number of times. Many of these materials are now lost to us. For example, did Enoch (Genesis 5:18, 21–24) really write books as tradition holds? What now exists as the books of Enoch are generally not believed to be written by him or to be part of the inspired Bible. However, Jude found enough truth in them to quote directly in Jude 1:14–15. Does this or any other pseudepigraphical, apocryphacal or non-Hebrew/Christian writings retain references from ancient works that give insight into prehistory? Probably, as the flood accounts from many cultures seem to point to.

Skeptics like to claim that these extra-biblical references are proof that writers of the biblical books often borrowed or copied from other contemporary or older writings. What these skeptics often gloss over is the fact that major doctrines or foundations of Christian theology are never "borrowed" from other writings. What we find borrowed (if they were indeed borrowed) are useful edicts that fit into the worldview of the Hebrews or Christians. Examples of this can be found in the Babylonian Code of Hammurabi, which contains similar ethical, legal and other rules that are found in the Bible.

Such biblical references to other works give us evidence of the ancient Hebrew's understanding of culture and language. We also see how they cut out the parts of surrounding cultures that were contrary to the laws given from God, but the Bible describes how they often failed to adhere to these laws and were influenced by practices of other nations.

As with flood accounts, history and mythology are also full of references to a "tree of life" or "tree of paradise," not unlike the tree found in Genesis. The Genesis accounts seem more like historical records, because they are not full of battling gods and other strange things written in human terms. As we have discussed, Genesis and other biblical books do not contradict science and contain knowledge unknown until centuries later. Some of those verses were scoffed at for years because they did not make sense, but no longer.

Skeptics will continue to claim the Bible "stole" and "borrowed" all sorts of beliefs from older works. As we have seen, such "borrowed" verses, or verses that reference other works, do not disprove the accuracy of the Bible. If anything, they provide authentication of the existence of the Hebrew people as contemporaries to other nations of the region. Also, notice that when skeptics claim that items such as the "tree of life" are not original to Christianity, ask them why they believe that. They will say, "Well, we have sources predating the Bible." I would respond, "So? Where are the direct links? Why are the context of these things so different in the Bible?" One could easily say that those older works were inspired by the same *historical* events that the Bible was, with those older works being corrupted by the ideas of man. In other words, the skeptic's reasoning does not hold up as proof of anything.

For the most part we can only speculate—and many speculate wildly—on which parts of ancient records truly describe prehistory and which are simply myth. Such faint echoes in history are all that remains of mankind's lost ancient heritage.

Appendix D

More on Young-Earth Creationism

Was there Death Before the Fall?

In four chapters the young-earth creationist view was detailed to some extent. Chapter 14 discussed at length various evidences that contradict the young-earth viewpoint. Some more young-earth issues will be addressed in this appendix. The first is the popular young-earth "proof" that I call the "death tautology" ("tautology" because it is often repeated over and over as an emotional plea).

The young-earth creationists claim that death did not enter the world until after man sinned (known as the fall of man). Hence, the billions of years of "death, struggle, disease and bloodshed" could not have happened. The death of animals, and especially people, is an obviously emotional occurrence. One will notice in young-earth writings on this subject that they tend to spend more time on the emotion rather than proving their theory. People get caught up in the emotion and do not bother to stop and ask, "Wait a second, is this true or am I letting emotion get the better of me?" Indeed, the whole argument is contingent on whether or not sin brought death into the world. Another important part of the debate would determine if death is inherently evil to begin with.

First, the Bible does not state that sin entered the world with Adam. Most seem to forget about Satan. When Satan and his minions rebelled against God, sin entered creation. Sin existed long before Adam arrived on the scene. This fact is bypassed on the way to quoting Romans 5:12. Often the "...and death [came] through sin..." part is held up without acknowledging the qualifying statement that follows it: "...in this way death came to all men..." The verse specifically relates man's sin to causing *man's* death, not all other death in nature. In other words, man's death was the result of his own choice and the Bible does not indicate that his sin caused all other death at that same moment. Since sin and evil existed before Adam, is that why the processes of death and decay existed before

127

man's creation? If that was the cause, then we have yet another support for death before Adam. But as we will soon see, death is not inherently evil, so sin or its evil purveyors could not be the sole cause (if they were a cause). In any case, since this physical death existed before Adam, then there is a good chance that the death described in Romans 5:12 refers to *spiritual* death.

At the very least, man must have died a spiritual death. Sin changed their lives and separated them from God. Christ came to die for humans to end this separation. Mankind's separation was spiritual, so a spiritual being had to end it. Both Romans 5:12 and 1 Corinthians 15:21 tell us specifically to who death came and why. Animals can be affected by our sin, but they cannot sin so there is no reason to believe that their natural death is caused by man's sin.

Death and Pain are Not Equivalent to Evil

Even plants "suffer" and experience "death," but some young-earth creationists have no problems with that. Nor do they have a problem with carnivorous plants that feed on animals. Did these plants suddenly appear after Adam's sin? According to the fossil record they did not. No death before Adam also violates the laws of physics. There is no life, or no work, without decay and death. In any given moment, cells are dying and food is decaying in our bodies so life may continue. Is this death and decay evil? If death was inherently evil, what of God who killed animals to clothe Adam and Eve and the deaths he caused throughout the Bible (in punishing people)? If death is inherently evil, then so is God.

Would not a caring creator prepare the world in the best possible way for man, the crown of creation? There are billions of tons of oil, coal, limestone, marble, topsoil and kerogen on Earth. All are valuable, and some necessary, for the maintaining and improvement of human life and all were created by decaying life. Would not the creator—knowing that man would sin by virtue of the fact the creator is outside of time and could see man's future—prepare the world accordingly? Was not the preparation of these materials good?

There is also the illogical claim that God did not create the laws of physics until *after* man sinned. Why would God create the laws *after* he created the universe? Those who believe death before Adam is evil explain the existence of carnivores by claiming that creation was already prepared to become meat-eaters since God knew what was coming. If this were true, the fossil record must be a deception, because it shows animals were already eating meat. Adam and Eve were not eating meat (which Genesis 1:29 and Genesis 9:3 seem to indicate) because it is healthier for people with their long life spans. With all the vegetation in Eden, meat would be unnecessary. When mankind's lifespan later decreased, he could eat reasonable amounts of meat without much worry (assuming meat is not the only food you eat, its good properties will outweigh the bad in our relatively short life spans).

When God gave man meat (Genesis 9:3) he did not say anything about changing the animals' diet. One could infer from this omission that the former guideline for animals (Genesis 1:30) was applicable only in Eden or was not completely forbidding carnivorous activity among animals. If Genesis 1:30 were forbidding all carnivorous activity, why does it only refer to the life types man would be interacting with inside Eden (land and airborne animals, see also Genesis 2:20) and not ocean dwelling creatures? To get really technical, read Genesis 1:29–30 again and notice how it is *not forbidding* anything, but seems to be a recommended guideline. Also, since God specifically *told* man he could eat meat in Genesis 9:3, here is another obvious point that the death of animals is not inherently evil.

Now consider how Genesis states Eve's childbearing pain was *increased* after the fall. This tells us two things: 1. There was pain before the fall; and 2. She may have had children before the fall. Also realize that pain is a defense mechanism, so it can not be a construct of evil.

People often ask why some people are taken from this world when they are. Granted, we do not have the larger perspective to see how everyone's life fits together, but consider that God may take people so they do not have to experience our corrupt and evil world anymore. On the other side of things, God limited the life spans of man so he could cause less death and destruction to each other. In other words, death serves larger purposes from a perspective beyond our own.

Once we strip away the emotion of death and look some of the *realities* of death, we find that the death tautology loses its strength. Not only through physical reality, but it does so through a careful reading of scripture as well.

Maybe the Universe only *Appears* Old

A popular young-earth theory often used (more in the past than now) is the claim that fossils, the universe, etc., only *appear* to be old. This seems to be the result of the reasoning, "O.K., we can't answer the tons of old-earth evidences, so maybe God made things appear this way for some reason." This appearance of age idea runs into a number of problems.

Most significantly is the fact that anything that *appears* to be old or has the *appearance* of age is, by definition, not really old. It only *appears* to be that way. Again, by definition, any such thing made to appear to be something that it is not, is to in fact be deceived by the person or being making it appear that way. Following this logic to its end would lead us to the conclusion that God is deceitful if he were in fact making things that appeared to be something other than what they are in actuality.

We cannot know the mind of God, but we know plenty concerning the nature of his character. Never is he irrational, untruthful or deceitful as made clear in verses such as Isaiah 45:19: "I have not spoken in secret, from somewhere in a land of darkness…" If you read the entire verse and the one preceding it, you will find that it is about God creating the universe in such a way that he and his message can be easily found. This seems contrary to suggestions that he fabricated the testimony of nature to appear to be something it is not. If God did do this, then how can we say the events in the Bible were not also fabricated? As you can see, appearance of age arguments open a Pandora's Box of problems.

Appearance of age is also tantamount to relative thinking. If something only appears to be old, how can we know anything of the past? One could then say we were created yesterday with memories and physical aging. In other words, appearance of age leads to some ridiculous lines of thought, which is why most have abandoned it as a possible young-earth support.

Appearance of age proponents may claim Adam appeared to be old since he was created as an adult instead of an infant. This reasoning does not pay close attention to the method of Adam's creation. Your age is calculated from the time you were born (we generally neglect the nine months in the womb for simplicity's sake). A year after Adam was created, he had only aged a year. By normal reckoning, he was only a year old. Adam was created as an adult so he was mature in the sense that he was not a child, not in the sense that he had already aged twenty years because he had not been alive that long. It would be as if you were born as a fully gown adult. Had this happened, would you show signs of aging at birth? Would you already have wrinkles, clogged arteries and worn joints? No, being created as an adult means just that: Adam was an adult from the get go and did not start aging until the next moment. Adam did not *appear* to be an adult, he *was* an adult. He only appears to be old from our perspective of being accustomed to people growing from an infant. From his perspective, Adam had just started aging!

Some will claim that when Jesus turned water into wine (John 2:1–11) that this was an example of appearance of age since we generally think of good wine as wine that has been aged. But did it have to be aged or is "new wine" acceptable? The Bible seems to indicate that in the culture of the day that new wine was not necessarily inferior and sometimes preferred (Proverbs 3:10, Joel 1:5, 3:18, Nehemiah 10:39, 13:12). In modern times we would more likely agree with Jesus' indication in Luke 5:39 that old wine is often better, but apparently old age is not necessary for good wine.

Lastly, some may ask, "If you cut a tree down in Eden, would it show growth rings?" It most certainly would. Genesis 2:8 states God "had" planted Eden, indicating that it had been there for awhile before Adam's creation. Recall from Chapter 14 that Adam was not created in Eden and time may have passed between his creation and placement in Eden. Also recall that day three of creation details vegetation-producing processes that require more than a few days.

What about those Dinosaurs?

People are fascinated by dinosaurs. These reptiles were very different from animals we are familiar with and they went extinct millions of years ago. Some seem almost like monsters, yet the fossil record proves they existed. The young-earth fascination with dinosaurs centers on "proving" they existed with man, hence indicating the world is not ancient. We already have discussed how the theory that a global flood deposited the fossil layers does not hold water. People who are familiar with paleontology or history realize that before the dinosaur fossil discoveries of the 1800s, no one had ever heard of the long lost reptiles. So what do young-earth creationists purport to be evidence of dinosaurs living alongside man?

The first evidence comes from mythology. Yes, you read that right. Creationists have a history of going through mythology, looking for references to dragons and other beasts—and even the Loch Ness monster—and pointing to these things as possible indicators of dinosaurs. Only recently did many creationists start agreeing with their critics that using mythology does not help one's scientific theories very well and some have stopped using these "proofs." Most have also abandoned using the "Loch Ness Monster" and similar frauds[1] and urban legends as evidences. Yet, as a sign of the poor dissemination of information in creationist circles, some still print these things as true. Even the long ago abandoned claim that human footprints were found alongside dinosaur tracks in the Paluxy River bed in Texas is resurrected quite often. Both secular and Christian researchers (including young-earthers) have concluded these prints are not human, but are dinosaur prints like the ones they are found with. Some reasons that the "human" prints are *not* human: 1. Too far apart to be human; 2. Most are too large; 3. Many show claw marks, etc.; and 4. Some of the "prints" are simply erosion patterns.[2]

In the light of the lack of tangible evidence, the focus is placed on the claim that the Bible refers to dinosaurs. Genesis does not specifically refer to dinosaurs when referring to the creation of animals. Some of the Hebrew could be *broadly* applied to include reptiles, however the context indicates that probably only animals that man was familiar with were intended. It is obviously not the point of Genesis to recount every object in creation.

Some people may also refer to older KJV translations of the Bible, which refer to a number of creatures as "dragons." More accurate translations have corrected these errors (and they are obvious translational errors from the original languages—Hebrew in the Old Testament and Greek in the New Testament) and have shown these "dragons" to be everything from jackals to snakes.

Now we come to the infamous account in Job 40 and 41, the "behemoth" and "leviathan" so often held up as absolute proof of dinosaurs and man living together. A careful review of these passages show that "literally" interpreting these words as dinosaurs is not literal at all. The first clue is the descriptions of the leviathan breathing fire and so forth. This sounds like figurative language describing the fear people had of this animal, after all what literal animal breathes fire? Also, the liberal use of "as" and "like" are dead giveaways of figurative speech.

Some Bibles footnote the behemoth as an elephant or hippopotamus. The leviathan sounds like another feared creature of Africa (such as in Egypt), the crocodile. Both the crocodile and hippo are extremely dangerous and have caused countless human deaths. They also fit the accounts in Job better than dinosaurs or mythical beasts.

Why an old Universe?

Is this old universe not a cruel and inefficient way of creating? We have already addressed this to some extent. First, it was noted that age is one of the constraints that limit where and when life exists and *if* it can exist. This is a function of the structure of the universe as it was created. "Cruelness" was also stripped away from death. We also saw how, over the life-span of Earth, different life-forms were used to create a biosphere capable of supporting mankind.

So essentially we have already answered the question, "Why an old universe?" in our other discussions. Apparently the creator felt this was the best way to prepare the world for mankind in such a way that the care and precision would someday become evident. To those who claim they could think of better ways to do the creating is to pretend to know the mind of a being beyond our existence. To have such a capability is irrational even to someone who does not believe in a higher being.

A more important question than "Why an old universe?" may be "Why are so many evidences for a designer and the accuracy of the Bible appearing *at this time in history?*"

The One-Sided Equation

Here is one last comment on young-earth "science" excerpted from Matthew S. Tiscareno's website "Is There Really Scientific Evidence for a Young Earth?" located at www.gps.caltech.edu/~tisco/yeclaimsbeta.html.

> A large class of "evidences" presented by young-Earth advocates involve measuring rates of various Earth processes, then attempting to extrapolate them backwards for millions of years. Generally, the purpose is to show that the process in question would build up to absurdity if it were allowed to continue through "evolutionary timescales." The fallacy of most claims of this type is a failure to recognize the importance of equilibrium. Most processes on Earth are in a state of balance, in which one process (such as erosion of the continents) is counteracted by others (such as emplacement of new continental material by volcanoes and tectonic uplift). Generally, processes on Earth do not build up without limit, because there is always another process that opposes the build-up, leading to the establishment of equilibrium. The method for dealing with young-Earth claims of this type is to look for the limiting process that imposes equilibrium. In some cases the balancing process has simply been overlooked, and the young-Earth claim is laid to rest by pointing it out…Examples of the One-Sided Equation Fallacy include Influx of Magma from Mantle to Form Crust, and Erosion of Sediment from Continents, Maximum Life of Comets, and Helium-4 in the Atmosphere.

APPENDIX E

Where did God Come from Anyway?

Even if intelligent design theory sounds compelling, you may still wonder as many do, "Who created God?" The answer may be a little unintuitive, but at the same time is not as hard as some think. Consider the nature of time discussed in Appendix A. It is something that is part of the properties of the universe that defines (or limits) the beings that live in it. Something bound by time is subject to cause and effect processes.

Every cause in this universe produces an effect. Each effect has a cause. Since cause and effect processes occur sequentially and in a forward direction on the timeline, they would not exist without time. An effect cannot precede a cause. The creation of something or someone only exists where there is time. However, what if something or someone was not bound by time? What if they were outside any dimension of time?

The universe, as we discussed, began at some point in the past with all of its properties and attributes (such as matter, energy, spacetime and so forth). Every creation requires something to set it off that is independent from the creation itself. The universe's formation included the creation of time, so whatever or whoever set it off was independent of time. If someone were not bound by time, what need would there be for his creation? If there is no time, then there are no cause and effect processes. An *uncaused* being is *impossible* in our universe, but where there is *no time* such a being *would not need a cause*.

It begins to make sense when we think this through, but it is still hard to visualize because we are used to our time-bound perspective. When we try to imagine no universe, what existed before it, an uncaused creator or how it can all be as it is, we eventually hit the brick wall of our mental capabilities. Sometimes we experience fear.

The following five arguments are more formalized arguments for the existence of God. The first three are related to what has already been discussed. These arguments are taken (and somewhat abbreviated) from "Twenty Arguments for the Existence of God" found in the *Handbook of Christian Apologetics* by Peter Kreeft and Ronald K. Tacelli. See this source for more detailed discussions on these proofs and answers to common objections on them.

1. Argument from Contingency

A.   If something exists, there must exist what it takes for that thing to exist.

B.   The universe—the collection of beings in space and time—exists.

C.   Therefore, there must exist what it takes for the universe to exist.

D.   What it takes for the universe to exist cannot exist within the universe or be bounded by space and time.

E.   Therefore, what it takes for the universe to exist must transcend both space and time.

2. The Kalam Argument (Kalam is Arabic for "speech")

A.   Whatever begins to exist has a cause for its coming into being.

B.   The universe began to exist.

C.   Therefore, the universe has a cause for its coming into being.

3. Design Argument

A.   The universe displays a staggering amount of intelligibility.

B.   Either this intelligible order is the product of chance or of intelligent design.

C.   Chance cannot produce such order.

D.   Therefore the universe is the product of intelligent design.

E.   Design comes only from a mind, a designer.

4. Argument from Truth

A. Our limited minds can discover eternal truths about being.

B. Truth properly resides in a mind.

C. But the human mind is not eternal.

D. Therefore there must exist an eternal mind in which these truths reside.

5. Argument from the Origin of the Idea of God

A. We have ideas of many things.

B. These ideas must arise from either ourselves or from things outside us.

C. One of these ideas is of an infinite, all-perfect being—God.

D. This idea could not have been caused by ourselves, because we know ourselves to be imperfect and no effect can be greater than its cause.

E. Therefore, the idea must have been caused by something outside us.

F. Only God, who has those qualities, could cause the idea we have of him.

G. Therefore God exists.

APPENDIX F

The Case for Christ

The existence of Christ and his death and resurrection are central to Christianity. Without the historical reality of these events, Christianity ceases to exist. Defense of these issues is the most critical apologetic issue in Christianity. This Appendix will briefly address this subject.

The Historicity of Christ

Few would deny Christ existed. Virtually all skeptics agree that he lived when and where Christians say he did. Denying he existed is like saying George Washington was a figment of our imagination. Yet there are those who still live in the Land of Make-Believe, so we will list some evidences here anyway.

1. To the dismay of skeptics, the Gospels were all written relatively close to the time of Jesus. Paul's epistles were written before the Gospels. So plenty of people who had seen, known or encountered Jesus were still alive to verify the writings or serve as sources.

2. Verses in Paul's epistles have been recognized as coming from early creeds that date within a few years (the first two), if not months, from the death of Jesus.

3. If writers were fabricating Jesus' existence, why would they make up events that were so contrary to the culture of the day? For example, they often showed Jesus giving women the respect rarely given to them in those times, treating them as equals to men.

4. Some other ancient, non-Christian sources that attest to Christ's existence include: Roman historian Cornelius Tacitus in *Annals*; Suetonius, chief secretary of Emperor Hadrian; Flavius Josephus, Jewish historian in *Antiquities*;

Julius Africanus in *Extant Writings* refers to the lost works of Thallus who describes the darkness and earthquake at the time of Christ's death; in Pliny the Younger's *Letters*; in letters from Emperors Trajan and Hadrian; Jewish documents the *Talmud* and the *Toledoth Jesu*; writings by Lucian, second century Greek satirist.

5.   The Gospels are extremely historically accurate in their details. The New Testament is also 99.5 percent free of textual discrepancies, which is very good considering there are thousands of ancient manuscripts. None of the discrepancies affect the doctrines of Christianity. Also consider, taking all these ancient writings together, there are more references to Jesus than most all other ancient individuals that we take for granted as being real.

Evidence for Christ's Death

Many claim Jesus did not die, but all such theories fail under scrutiny. Worse, most are just speculations or ad hoc explanations. Here we will look at some compelling evidences for Christ's death.

1.   Some of the ancient non-Christian sources above attest to Christ's execution. At minimum, hundreds would have been witness to the spectacle surrounding the event. In other words, there were many first-hand eyewitnesses.

2.   Archaeology also confirms death by crucifixion and the use of nails to "hang" the person.

3.   The Romans were quite proficient in crucifying people and did so to many thousands. Crucifixion was very brutal, the term *excruciating* comes from "out of the cross." To think the Romans messed up on Jesus does not make sense. They also had ways and motivations to ensure the person was dead (see three points below).

4.   Hanging on a cross, one would die of asphyxiation. The hanging position applied pressures to the diaphragm (the muscle that controls breathing) that would require the person to push up with his feat to relieve the pressure and to breathe. This is why they would break the legs of crucified individuals to speed up their deaths. The soldiers did this to the two hung with Jesus, but not him because they found him already dead.

5. The soldiers guarding Jesus would have paid for their life if they did not make sure Jesus was dead before taking him down off the cross.

6. Perhaps the most telling evidence is when John describes "blood and water" coming from Jesus' side when the soldier stabbed him with a spear to confirm he was dead. John probably had no idea what this was, but modern science confirms that this "water" was fluid known as the pericardial and pleural effusion surrounding the heart.

The Empty Tomb

The evidence that Jesus was killed negates the many fantastic theories of Jesus letting himself out of the tomb, or escaping to live in other countries, getting married and so on. Here we will address why the placement in the tomb and its subsequent emptiness was not some sort of fraud or hoax.

1. Once again there is the witness argument. People would have seen him placed in the tomb.

2. The Bible relates that the tomb belonged to Joseph of Arimathea, a member of the Sanhedrin. Given that the Sanhedrin was partly responsible for Jesus' death, it is curious that the Gospel writers would claim Joseph of Arimathea had donated his tomb to Jesus if they were writing an imaginary tale. Many people would have been in the position to dispute their claim if it were not true.

3. Given the culture of the day, if the Gospel writers were creating fiction, the last people they would have finding the tomb empty were women. Yet this is who made the discovery.

4. It would be hard to steal the body, which was under guard to prevent theft. The government did not want Jesus' followers to incite further unrest by taking the body. Moving the stone from the tomb is no simple task either.

5. The fact is that the earliest skeptical claims do not deny the tomb was empty, but claim the body was stolen. Yet they, nor any skeptics since, have provided any historical evidence. In other words they admitted the tomb was empty, but did not have an adequate explanation for why it was empty.

Appearances of Christ

If Jesus was really dead and the tomb really found empty, as the evidence suggests, then how does one explain his post-death appearances?

1.   Again, it would be crazy for writers in that culture to claim that women were the first to see the resurrected Jesus.

2.   Again, the writings of Paul reference materials from shortly after Christ's death. He was dealing with primary sources, not years of story telling or myth-making. Indeed, that can be applied to all of these categories: There was not time for legends and myths to form and there were too many people alive at the time that could dispute such claims.

3.   Paul was in fact a great persecutor of Christians until he encountered Christ *after* his death. Paul would go on to suffer the same types of persecutions. So it seems curious that he would make up his writings attesting to Christ's life, death and resurrection.

4.   Hallucinations do not provide an explanation. They are a highly individualistic experience. Only particular people are susceptible to such things. Often it is only at particular times and places. A great variety of people claimed to have seen Christ after he died at different times and places. This makes it hard to claim that they were all hallucinating.

5.   Again it is important to reiterate that there were many witnesses who believed that they had seen and interacted with Christ after he died. These events often happened in public places, not hidden away in a basement or cave. They made such claims in the face of fierce and often deadly persecution. Many died for their beliefs. This is different from Muslims who die for their belief that Mohammed received revelations from God. They did *not* witness this happening, so they could be wrong. Early Christians based their beliefs on the fact that they *did* witness Christ's death and resurrection, so they had no reason to believe they were wrong.

Scholars have committed volumes to the defense of Christ, his life, death and resurrection. Some of the sources listed for this Appendix go into much more depth and detail on the points that we have only briefly visited. Perhaps the most important clue can be found by comparing these things with other events in antiquity. By the standards we apply to these other events, the events surrounding Christ's life, death and resurrection are just as historical, even more so than much of what we accept as history.

> True, the discovery of the empty tomb is differently described by the various gospels, but if we apply the same sort of criteria that we would apply to any other ancient literary sources, the evidence is firm and plausible enough to necessitate the conclusion that the tomb was, indeed, found empty.—skeptical historian Michael Grant[1]

> Even the more skeptical historians agree that for primitive Christianity...the resurrection of Jesus from the dead was a real event in history, the very foundation of faith, and not a mythical idea arising out of the creative imagination of believers.—theologian and historian Carl Braaten[2]

> The [post-resurrection] appearances of Jesus are as well authenticated as anything in antiquity...There can be no rational doubt that they occurred...—theologian Michael Green[3]

Notes and Bibliography

For easier study, and to the dismay of grammatical purists, I have combined the Notes and Bibliography section. Instead of complicating and cluttering the sources and references for chapters on science issues with hundreds of scientific articles, I mostly reference secondary sources. So if a subject interests you, these sources detail at length these science topics in understandable fashion. If you want to research at a deeper level, many of these sources list extensive scientific citations. Internet links were up-to-date at publication time.

Simply because a resource is listed here does not mean the author of this book agrees with every point within. Nor does it mean you should ignore the Maxims and Rules while reading them.

Epigraph

The quote from King Solomon is recorded in Proverbs 1:20–21; 3:13–14, 21–22; 4:13; 8:36 from the New International Version (NIV) translation. The NIV is used as the main reference translation for this book unless noted otherwise.

Introduction

1. Phillip E. Johnson, *The Wedge of Truth* (Downers Grove, Illinois: InterVarsity Press, 2000), p. 9.

Chapter 1: Is There Absolute Truth?

1. Dean C. Halverson, editor, *The Compact Guide to World Religions* (Minneapolis, Minnesota: Bethany House, 1996), p. 30.

2. Kenneth D. Boa and Robert M. Bowman, Jr., *20 Compelling Evidences That God Exists* (Tulsa, Oklahoma: River Oak, 2002). On page 300 the authors describe the defining of atheism this way:

    "Atheists sometimes complain that the word *atheism* means merely the lack of a belief in God, not the belief that no God exists. This isn't correct: the word *atheism*, like similar formed words *monotheism, polytheism,* and *pantheism,* refers to a particular view about the world and its relation to the divine, if any."

Chuck Colson and Nancy Pearcey, *How Now Shall We Live?* Wheaton, Illinois: Tyndale House, 1999. This is a very detailed and authoritative work comparing the worldviews of theism and naturalism.

Phillip E. Johnson, *Reason in the Balance: The Case Against Naturalism in Science, Law & Education.* Downers Grove, Illinois: InterVarsity Press, 1995.

R.C. Sproul, *Not a Chance: The Myth of Chance in Modern Science & Cosmology.* Grand Rapids, Michigan: Baker Books, 1994.

William D. Watkins, *The New Absolutes.* Minneapolis, Minnesota: Bethany House, 1996.

Chapter 2: Critical Thinking Tool Kit

1. William A. Dembski & Jay Wesley Richards, editors, *Unapologetic Apologetics* (Downers Grove, Illinois: InterVaristy Press, 2001), p. 86.

2. Phillip E. Johnson, *The Wedge of Truth* (Downers Grove, Illinois: InterVarsity Press, 2000), pp. 9–10.

3. Hugh Ross, Kenneth Samples and Mark Clark, *Lights in the Sky and Little Green Men* (Colorado Springs, Colorado: NavPress, 2002), pp. 137–138.

Carl Sagan, *The Demon-Haunted World: Science as a Candle in the Dark.* New York, NY: Ballantine Books, 1996. Sagan debunks a lot pseudoscience in this book. However, when it came to his own beliefs and other closely held theories he ignored the clear thinking, which he presented as a "Baloney Detector." One such "Sagan Paradigm" is that Earth is just a random speck among the stars, unconnected to the universe at large. See Chapter 6 for more on this fallacy. Phillip E. Johnson briefly addresses other issues with Sagan in his *Defeating Darwinism by Opening Minds* (see below).

J.P. Moreland, *Love Your God With All Your Mind: The Role of Reason in the Life of the Soul.* Colorado Springs, Colorado: NavPress, 1997.

Norman L. Giesler and Ronald M. Brooks, *Come, Let Us Reason: An Introduction to Logical Thinking.* Grand Rapids, Michigan: Baker Book House, 1998.

Phillip E. Johnson, *Defeating Darwinism by Opening Minds.* Downers Grove, Illinois: InterVarsity Press, 1997.

Rory Coker, "Distinguishing Science and Pseudoscience." Website at www.quackwatch.org/01QuackeryRelatedTopics/pseudo.html.

Chapter 3 and Chapter 4: Changing History Without a Time Machine & Ghosts of History

Chuck Colson with Anne Morse, *Burden of Truth: Defending Truth in an Age of Unbelief.* Wheaton, Illinois: Tyndale House, 1997.

Chuck Colson with Nancy R. Pearcey, *A Dance with Deception: Revealing the Truth Behind the Headlines.* Dallas, Texas: Word, 1993.

Chuck Colson and Nancy Pearcey, *How Now Shall We Live?* Wheaton, Illinois: Tyndale House, 1999.

Craig L. Symonds, *History of the Battle of Gettysburg.* New York, NY: Byron Preiss, 2001.

Douglas Kennedy, "Texas History Gets New Mexican Twist," FoxNews.com (20 May 2002): On-line at www.foxnews.com/story/0,2933,53968,00.html.

John Keegan, *The First World War.* New York, NY: Knopf, 1999.

Joseph Dahmus, *A History of the Middle Ages.* New York, NY: Barnes & Nobles Books, 1995.

Mark Bowden, *Black Hawk Down: A Story of Modern War.* New York, NY: Penguin, 2000.

Martin Gilbert, *The First World War: A Complete History.* New York, NY: Owl Books, 1994.

Noah Andre Trudeau, *Gettysburg: A Testing of Courage.* New York, NY: Harper-Collins, 2002.

Peter James and Nick Thorpe, *Ancient Mysteries.* New York, NY: Ballantine Books, 1999.

Randall Price, *Unholy War: America, Israel and Radical Islam.* Eugene, Oregon: Harvest House, 2001.

Robert Leckie, *George Washington's War.* New York, NY: HarperPerenial, 1993.

Robin Moore, *The Hunt For Bin Laden, Task Force Dagger: On the Ground with Special Forces in Afghanistan.* New York, NY: Random House, 2003.

Susan Wels, *Pearl Harbor: America's Darkest Day.* Sand Diego, CA: Tehabi Books, 2001.

Tim Newark, *Turing the Tide of War: 50 Battles that Changed the Course of Modern History.* Heron Quays, London: Hamlyn, 2001.

William D. Watkins, *The New Absolutes.* Minneapolis, Minnesota: Bethany House, 1996.

Chapter 5: Abducting Your Mind

1. Hugh Ross, Kenneth Samples and Mark Clark, *Lights in the Sky and Little Green Men* (Colorado Springs, Colorado: NavPress, 2002), pp. 65–71. Gives a detailed review of the UFO phenomenon, the government's part (or supposed part) in all this and a thorough look at the nature of RUFOs.

2. André Kole and Jerry MacGregor, *Mind Games: Exposing Today's Psychics, Frauds, and False Spiritual Phenomena* (Eugene, Oregon: Harvest House, 1998), pp. 157–161.

3. Ibid., pp. 114–115.

4. Tony Allan, *Prophecies* (London, England: Thorsans, 2002), pp. 128–129.

5. Accounts of people who have premonitions of impeding doom are not uncommon. Abraham Lincoln foresaw his death three days earlier in a dream. Accounts of people not getting on a plane or ship because of a bad feeling are often reported surrounding tragic events such as the sinking of the *Titanic*. These are not technically prophecies but are related (pages 58–71 of the source in Note 4 discusses premonitions in a little more detail).

    Many have made a case for the accuracy of biblical prophets. In "Fulfilled Prophecy: Evidence for the Reliability of the Bible" (on-line at www. reasons.org/resources/apologetics/prophecy.shtml?main), Hugh Ross does a probability study of thirteen prophecies (out of over 2000) in the Bible. Each of these predictions has come true, however, the chance of them having been fulfilled is 1 in 10^{138}. Discussing biblical prophecy is beyond the scope (and space) of this book. More introductions to this subject can be found in Richard Deem's articles "Prophecies of Jesus Christ as Messiah" and "Prophetic Evidence" at www.evidence.info/theology/prophchr.html and www. evidence.info/authenticity/bibletru.html#Prophetic, respectively.

Alex Boese, *The Museum of Hoaxes*. New York, NY: Dutton, 2002. See also the author's website at www.museumofhoaxes.com. A similar site can be found at www.truthorfiction.com.

"BBC 'proves' Nessie Does Not Exist." Article for *BBC News* on-line at news.bbc.co.uk/2/hi/science/nature/3096839.stm, July 27, 2003.

Carl Sagan, *The Demon-Haunted World: Science as a Candle in the Dark*. New York, NY: Ballantine Books, 1996.

Kenneth Samples, "Alien Encounters Fail the Test." *Facts for Faith* Issue 6 (Q2 2001): pp. 52–59. Also available on-line at www.reasons.org/resources/fff/2001issue06/index.shtml?main#alien_encounters.

Philip Plait, *Bad Astronomy*. New York, NY: Wiley, 2002. This book covers some of the same science myths and others in a lot more detail. Overall it is a fairly good resource. Based on Plait's website www.badastronomy.com.

Robert L. Park, "The Seven Warning Signs of Bogus Science." *The Chronicle Review* Volume 49, Issue 21 (January 31, 2003): p. B20. Also on-line at chronicle.com/free/v49/i21/21b02001.htm.

Steven Milloy, www.junkscience.com. Studies and articles on "junk science" items and issues. Also authors some books on the subject. A similar site dedicated to medical issues is found at www.quackwatch.org.

Tim Parker, "Picturing Cydonia." *Ad Astra* (July/August 1998): pp. 30–32.

Chapter 6: What About Those Aliens?

1. Miroslav Verner, *The Pyramids: The Mystery, Culture, and Science of Egypt's Great Monuments* (New York, NY: Grove Press, 2001). Detailed look at ancient Egyptian monuments. No aliens or lost super intelligent human civilizations required.

2. Anthony Aveni, *Stairways to the Stars: Skywatching in Three Great Ancient Cultures* (New York, NY: John Wiley & Sons, 1997). Discusses Stonehenge and Mayan and Inca structures and their possible uses in astronomy. Nothing real alien here.

3. Hugh Ross, *The Creator and the Cosmos*, 3rd ed. Rev. (Colorado Springs, Colorado: NavPress, 2001), pp. 195–198. On pages 153–154, Ross mentions that the best manmade creation is only "accurate to one part in 10^{23}." Apparently, the design found in the universe was created by intelligence far greater than our own.

4. Hugh Ross, Kenneth Samples and Mark Clark, *Lights in the Sky and Little Green Men* (Colorado Springs, Colorado: NavPress, 2002), p. 39. Hugh Ross puts the probability (1 chance in 10^{174}) in perspective by writing, "…the universe contains only 10^{79} protons and neutrons, and every reader has a much higher probability of being killed in the next second by a failure in the second law of thermodynamics (about one chance in 10^{80})."

André Kole and Jerry MacGregor, *Mind Games: Exposing Today's Psychics, Frauds, and False Spiritual Phenomena*. Eugene, Oregon: Harvest House, 1998.

Bijan Nemati, "The Search for Life on Other Planets." *Facts for Faith* Issue 4 (Q4 2000): pp. 22–31.

Guillermo Gonzalez, "Rare Sun." *Facts for Faith* Issue 9 (Q2 2002): pp. 14–21. Also available on-line at www.reasons.org/resources/fff/2002issue09/index.shtml?main#rare_sun.

Hugh Ross, "Aliens from Another World?" *Facts for Faith* Issue 6 (Q2 2001): pp. 24–32. Also available on-line at www.reasons.org/resources/fff/2001issue06/index.shtml?main#aliens_from_another_world.

Hugh Ross, "Exotic Life Sites: The Feasibility of Far-Out Habitats." *Facts for Faith* Issue 7 (Q4 2001): pp. 20–25. Also available on-line at www.reasons.org/resources/fff/2001issue07/index.shtml?main#exotic_life.

Mark A. Garlick, "No Place Like Zone." *Astronomy* Volume 30, Number 8 (August 2002): pp. 44–51.

Michael A. Corey, *The God Hypothesis: Discovering Design in Our Just Right Goldilocks Universe.* Rowman & Littlefield, 2002.

Michael J. Denton, *Nature's Destiny: How the Laws of Biology Reveal Purpose in the Universe.* New York, NY: Free Press, 1998.

Peter Douglas Ward and Donald Brownlee, *Rare Earth: Why Complex Life is Uncommon in the Universe.* New York, NY: Copernicus, 2000.

Some related resources can be found in the Notes for Chapter 15.

Chapter 7: Figuring out Science

1. Hugh Ross, *Beyond the Cosmos*, 2rd ed. Rev. (Colorado Springs, Colorado: NavPress, 1999), p. 29.

Arthur Upgreen, *Night Has a Thousand Eyes.* Cambridge, Massachusetts: Perseus, 1998.

Bob Berman, "Bad Moon Rising." *Astronomy* Volume 30, Number 9 (September 2002): pp. 96–97.

Bob Berman, *Secrets of the Night Sky.* New York, NY: HarperPerennial, 1997. Sky lore and naked eye stargazing.

Brian Greene, *The Elegant Universe.* New York, NY: Vintage Books, 1999. An excellent journey through modern physics including Einstein's discoveries, the nature of time, hidden dimensions and superstring theory.

Curt Suplee, *Physics in the 20th Century*. New York, NY: Abrams, 1999. A detailed and well-illustrated look at the amazing advances in physics in the Twentieth Century.

J. Richard Gott, *Time Travel in Einstein's Universe*. New York, NY: Houghton Mifflin, 2002. The author does a good job describing the intricacies of time, before going off into his own speculative theories.

Philip Plait, *Bad Astronomy*. New York, NY: Wiley, 2002.

R. Mike Mullane. *Do Your Ears Pop in Space?* New York, NY: Wiley, 1997. Former shuttle astronaut Mullane answers hundreds of questions about spaceflight.

Tony Rothman, *Instant Physics*. New York, NY: Byron Preis, 1995. A great introduction or review to physics in an accessible format.

William A. Gutsch. *1001 Things Everyone Should Know About the Universe*. New York, NY: Doubleday, 1998. An accessible and thorough journey through astronomy.

Chapter 8: Environmental Protection or Ecoterrorism?

1. "Dumb and Dumber: What animal-rights activists say about Sept. 11." *Outdoor Life* Volume 209, Number 3 (May 2002): p. 17.

Chapter 9: Testing All Things

1. Philip Plait, *Bad Astronomy* (New York, NY: Wiley, 2002), p. 190.

2. Hugh Ross, *The Genesis Question*, 2nd ed. (Colorado Springs, Colorado: NavPress, 2001), pp. 195–197. This book also thoroughly details reconciling the much maligned first eleven chapters of Genesis with science.

3. Guy Consolmagno, *Brother Astronomer* (New York, NY: McGraw-Hill, 2000), p. 85. The author also gives a detailed account of the "Galileo Affair."

4. Hugh Ross, *The Fingerprint of God*, 2nd ed. (Orange, California: Promise, 1991), p. 21.

5.   William A. Dembski & Jay Wesley Richards, editors, *Unapologetic Apologetics* (Downers Grove, Illinois: InterVaristy Press, 2001), p. 38.

Dembski writes "…(Protestant, Roman Catholic and Eastern Orthodox confessions traditionally have been united on these cases): To the physical core of the Christian faith belong the virgin birth, the crucifixion and the resurrection. To the theoretical core belong the incarnation, the redemption through Christ and the Trinity. To the regulative core belong the reliability of Scripture, the preeminence of Christ and a commitment to truth."

This is not meant to be exhaustive, but these are some of the beliefs orthodox Christian denominations share. The "theoretical core" explains the "physical core" and the "regulative core" govern how these beliefs are applied in practice. Dembski makes these distinctions and discusses them on pages 35–41. It is against these cores that all traditions and beliefs need to be moderated by or tested against within Christianity.

6.   Chuck Colson and Nancy Pearcey, *How Now Shall We Live?* (Wheaton, Illinois: Tyndale House, 1999), p. 270.

7.   André Kole and Jerry MacGregor, *Mind Games: Exposing Today's Psychics, Frauds, and False Spiritual Phenomena* (Eugene, Oregon: Harvest House, 1998), p. 118.

8.   Ibid., p. 115.

9.   Ibid., pp. 107 and 116. See also Richard Deem, "Cults of Christianity: Mormonism" at www.evidence.info/cults/index.html for more information on Mormonism.

10. Robert C. Newman, "Joshua's Long Day and the NASA Computers: Is the Story True?" Website at www.reasons.org/resources/apologetics/joshualongday.shtml.

11. David Bloomberg, "The Incredible Mysteries of Sun Pictures." Website at www.talkorigins.org/faqs/ark-hoax/sun.html. See also Hugh Ross, "The Unsinkable Search for Noah's Ark." Website at www.reasons.org/resources/faf/93q1faf/93q1nark.shtml?main.

12. Edward B. Davis, "A Modern Jonah." Website at www.reasons.org/resources/apologetics/jonah.shtml.

13. Rich Buhler, "Background to the Drilling to Hell Story." Website at www.truthorfiction.com/rumors/drilltohellfacts.htm. "TruthOrFiction.com is a web site where Internet users can quickly and easily get information about eRumors, warnings, offers, requests for help, myths, hoaxes, virus warnings, and humorous or inspirational stories that are circulated by email." A similar site can be found at www.museumofhoaxes.com.

14. Randall Ingermanson, *Who Wrote the Bible Code? A Physicist Probes the Current Controversy.* Colorado Springs, Colorado: Waterbook, 1999. The author maintains additional information and appendices to his book on-line at www.rsingermanson.com. More on the Bible codes can be found at www.reasons.org/resources/apologetics/biblecode.shtml?main.

15. William A. Dembski & Jay Wesley Richards, editors, *Unapologetic Apologetics* (Downers Grove, Illinois: InterVaristy Press, 2001), p. 43.

Some General Apologetics Resources:

Answers in Action, www.answers.org.

Cliffe Knechtle, *Give me an Answer That Satisfies my Heart and my Mind.* Downers Grove, Illinois: InterVarsity Press, 1986.

Chuck Colson, *Answers to your Kids' Questions.* Wheaton, Illinois: Tyndale House, 2000.

Dean C. Halverson, editor, *The Compact Guide to World Religions.* Minneapolis, Minnesota: Bethany House, 1996.

Gary R. Habermas, *The Historical Jesus: Ancient Evidence for the Life of Christ.* Joplin, Missouri: College Press, 1996.

Gordon D. Fee and Douglas Stuart, *How to Read the Bible for all its Worth.* Grand Rapids, Michigan: Zondervan, 1993.

Gregory Koukl, "Arguing is a Virtue." Website at www.str.org/free/commentaries/apologetics/records/arguing.htm.

Gregory Koukl, "Testing Religious Truth Claims." Website at www.str.org/free/commentaries/theology/testingr.htm.

Hugh Ross, "Who's Right? Who's Wrong? Guidelines of Christian Scholarship." Website at www.reasons.org/resources/apologetics/christiansch.shtml?main.

J.P. Moreland, *Love Your God With All Your Mind: The Role of Reason in the Life of the Soul.* Colorado Springs, Colorado: NavPress, 1997. An excellent look at the loss of the mind in Christianity.

Kenneth D. Boa and Robert M. Bowman, Jr., *Faith has its Reasons: An Integrative Approach to Defending Christianity.* Colorado Springs, Colorado: NavPress, 2001.

Kenneth D. Boa and Robert M. Bowman, Jr., *20 Compelling Evidences That God Exists.* Tulsa, Oklahoma: River Oak, 2002. Covers both science and nonscience issues.

Lee Strobel, *The Case for Christ.* Grand Rapids, Michigan: Zondervan, 1998.

Lee Strobel, *The Case for Faith.* Grand Rapids, Michigan: Zondervan, 2000. Covers both science and nonscience issues.

Norman L. Geisler and Abdul Saleeb, *Answering Islam: The Crescent in Light of the Cross*, 2nd ed. Grand Rapids, Michigan: Baker Book House, 2002.

Norman L. Geisler and Ronald M. Brooks, *When Skeptics Ask.* Wheaton, Illinois: Victor Books, 1990.

Peter Kreeft & Ronald K. Tacelli, *Handbook of Christian Apologetics.* Downers Grove, Illinois: InterVaristy Press, 1994.

Randall Price, *The Stones Cry Out.* Eugene Oregon: Harvest House, 1997.

Stand to Reason, www.str.org.

Sword and Spirit, www.swordandspirit.com.

Some Science Apologetics Resources (more in the Notes for Chapter 13, 14 and 15):

Chuck Colson and Nancy Pearcey, *Developing a Christian Worldview of Science and Evolution.* Wheaton, Illinois: Tyndale House, 2001.

Colin J. Humphreys, *The Miracles of Exodus: A Scientist's Discovery of the Extraordinary Natural Causes of the Biblical Stories*. New York, NY: HarperSanFrancisco, 2003. Extensive look at the issues raised by events recorded in the Book of Exodus.

Don Stoner, *A New Look at an Old Earth: Resolving the Conflict Between the Bible & Science*. Eugene, Oregon: Harvest House, 1997 edition. On-line version at answers.org/newlook/NEWLOOK.HTM#New.

Does God Exist? www.doesgodexist.org.

Evidence for God, www.evidence.info.

Fred Heeren, *Show me God*, Rev. ed. Wheeling, Illinois: Day Star, 1997.

Hugh Ross, *Beyond the Cosmos*, 2rd ed. Rev. Colorado Springs, Colorado: Nav-Press, 1999.

Hugh Ross, *Creation and Time: A Biblical and Scientific Perspective on the Creation-Date Controversy*. Colorado Springs, Colorado: NavPress, 1994.

Hugh Ross, *The Creator and the Cosmos*, 3rd ed. Rev. Colorado Springs, Colorado: NavPress, 2001.

Hugh Ross, *The Genesis Question*, 2nd ed. Colorado Springs, Colorado: NavPress, 2001. Extensive look at the issues raised by events recorded in the Book of Genesis.

Hugh Ross, Kenneth Samples and Mark Clark, *Lights in the Sky and Little Green Men*. Colorado Springs, Colorado: NavPress, 2002.

Jim Schicantano, *The Theory of Creation*. New York, NY: Writers Club Press, 2001.

Kenneth Richard Samples, "The Historic Alliance of Christianity and Science." Website at www.reasons.org/resources/apologetics/christianscience.shtml?main.

R.C. Sproul, *Not a Chance: The Myth of Chance in Modern Science & Cosmology*. Grand Rapids, Michigan: Baker Books, 1994.

Reasons to Believe, www.reasons.org.

Chapter 10: Getting a Bang out of the Universe

1. A. Vibert Douglas, "Forty Minutes with Einstein," *Journal of the Royal Astronomical Society of Canada 50* (1956), p. 100.

2. "'Big Bang' Critic Passes Away." *Answers Update* (V8, N10: Oct 2001), p. 4.

3. From a letter by Matthew S. Tiscareno, a Ph.D. candidate in Planetary Science at the University of Arizona. He also authors the "Is There Really Scientific Evidence for a Young Earth?" website at www.gps.caltech.edu/~tisco/yeclaimsbeta.html.

4. Fred Hoyle, "The Universe: Past and Present Reflections," *Annual Review of Astronomy and Astrophysics 20* (1982), p. 16.

Some Fairly Recent and Accessible Resources on the Big Bang and Related Matters:

Brian Greene, *The Elegant Universe.* New York, NY: Vintage Books, 1999.

Edward Witten, "Universe on a String." *Astronomy* Volume 30, Number 6 (June 2002): pp. 40–45.

Fred Heeren, *Show me God*, Rev. ed. Wheeling, Illinois: Day Star, 1997. Contains a good introduction to the big bang, its history and implications as well as discussing the implications for God from Twentieth Century astrophysics.

Hugh Ross, "A Beginner's—and Expert's—Guide to the Big Bang." *Facts for Faith* Issue 3 (Q3 2000): pp. 14–32.

Hugh Ross, "Facing up to Big Bang Challenges." *Facts for Faith* Issue 5 (Q1 2001): pp. 42–53. Also available on-line at www.reasons.org/resources/fff/2001issue5/index.shtml?main#big_bang_challenges.

Hugh Ross, "Predictive Power: Confirming the Cosmic Creation." *Facts for Faith* Issue 9 (Q2 2002): pp. 32–39. Also available on-line at www.reasons.org/resources/fff/2002issue9/index.shtml?main#predictive_power.

Hugh Ross, *The Creator and the Cosmos*, 3rd ed. Rev. Colorado Springs, Colorado: NavPress, 2001. Astronomer Ross details the evidences (at least 30) for big bang science and its theological implications at a very understandable level. Some additional updates and additions can be found at www.reasons.org/resources/apologetics/big_bang_evidences.shtml?main.

Jim Sweitzer, "Do You Believe in the Big Bang?" *Astronomy* Volume 30, Number 12 (December 2002): pp. 34–39.

Steve Nadis, "Cosmic Inflation Comes of Age." *Astronomy* Volume 30, Number 4 (April 2002): pp. 28–32.

Steve Nadis, "When Branes Collide." *Astronomy* Volume 30, Number 5 (May 2002): pp. 34–39.

Chapter 11: Flood of Beliefs

1. Hugh Ross, *The Genesis Question*, 2nd ed. (Colorado Springs, Colorado: NavPress, 2001). Probably one of the best looks at the "global vs. local flood" issue in Chapters 17–19.

    Ross spends Chapter 17 discussing how the punishment of man is always based on the "extent of sin's damage" and what the sin has actually defiled. On page 140, he describes how defilement from sin spreads from the sinner (Romans 7:8–11), to his progeny (Exodus 20:5), to his soulish animals (Joshua 6:21), to his material goods (Numbers 16:23–33) and then to his lands (Leviticus 18:24–28).

Alan Hayward, *Creation and Evolution: Rethinking the Evidence from Science and the Bible*. Minneapolis, Minnesota: Bethany House, 1985, (1995 printing).

Bill T. Arnold, *Encountering the Book of Genesis*. Grand Rapids, Michigan: Baker Books, 1998.

Don Stoner, *A New Look at an Old Earth: Resolving the Conflict Between the Bible & Science*. Eugene, Oregon: Harvest House, 1997 edition.

John M. Clayton, "A Visit to Mt. St. Helens." Website at www.doesgodexist.org/MarApr01/AVisitToMtStHelens.html.

Karen Bartelt, "A Visit to the Institute for Creation Research." Website at www.talkorigins.org/faqs/icr-visit/bartelt1.html. While the author occasionally delves into evolutionary pseudoscience, her discussions on bad young-earth science relating to geology, Mt. St. Helens and the Grand Canyon are informative.

Richard Deem, "The Genesis Flood: Why the Bible Says It Must be Local." Website at www.evidence.info/creation/localflood.html.

Ronald L. Numbers, *The Creationists: The Evolution of Scientific Creationism.* Berkley, California: University of California Press, 1992.

Steve Sarigianis, "Noah's Flood: A Bird's Eye View." *Facts for Faith* Issue 10 (Q3 2002): pp. 16–25. Also available on-line at www.reasons.org/resources/fff/2002issue10/index.shtml?main#noahs_flood.

Chapter 12: Reasonable or Skewed Science: The Evolution vs. Creation Debate

1. See also: Daniel 5 where Belshazzar is said to be the son of Nebuchadnezzar, where in fact he was the son of Nabonidus and Ruth 4:17 where Obed is "born to Naomi" when Nomi is in fact the mother-in-law of Obed's mother. More examples of gaps (or "telescoping") in the genealogies can be found in comparing Matthew 1:11 to 2 Chronicles 36:1–9, Luke 3:35–36 to Genesis 10:24, 11:12 and 1 Samuel 16:10–13 to 1 Chronicles 7:13–15. These items and more are discussed in John Millam's article "The Genesis Genealogies" on-line at www.geocities.com/darrickdean/gengene.html.

    Colin J. Humphreys in his book, *The Miracles of Exodus: A Scientist's Discovery of the Extraordinary Natural Causes of the Biblical Stories* (New York, NY: HarperSanFrancisco, 2003), also discusses some important theories in interpreting the differences between the genealogies found in Ezra 7:1–5 and 1 Chronicles 6:3–13. This may be important in reconciling the date of the Exodus with Egyptian chronologies. See Chapter 3 in his book.

2. The incompleteness of the genealogies has long been recognized and is the only way to resolve apparent problems in the text and reconcile the text with scientific finds. Dr. William Henry Green's article "Primeval Chronology," written in the April 1890 issue of *Bibliotheca Sacra* comments at length on

this issue. Available on-line as a PDF file at lordibelieve.org/time/ WHGreen.PDF. Here is a brief excerpt:

"The structure of the genealogies in Genesis 5 and 11 also favors the belief that they do not register all the names in these respective lines of descent. Their regularity seems to indicate intentional arrangement. Each genealogy includes ten names, Noah being the tenth from Adam, and Terah the tenth from Noah...Now this adjustment of the genealogy in Matthew 1 into three periods of fourteen generations each is brought about by dropping the requisite number of names, it seems in the highest degree probable that the symmetry of these primitive genealogies is artificial rather than natural. It is much more likely that this definite number of names fitting into a regular scheme has been selected as sufficiently representing the periods to which they belong, than that all these striking numerical coincidences should have happened to occur in these successive instances.

"It may further be added that if the genealogy in Chapter 11 is complete, Peleg, who marks the entrance of a new period, died while all his ancestors from Noah onward were still living. Indeed Shem, Arphaxad, Selah, and Eber must all have outlived not only Peleg, but all the generations following as far as and including Terah. The whole impression of the narrative in Abraham's days is that the Flood was an event long since past, and that the actors in it had passed away ages before. And yet if a chronology is to be [literally] constructed out of this genealogy, Noah was for fifty-eight years the contemporary of Abraham, and Shem actually survived him thirty-five years, provided 11:26 is to be taken in its natural sense, that Abraham was born in Terah's seventieth year. This conclusion is well-nigh incredible. The calculation which leads to such a result, must proceed upon a wrong assumption.

"On these various grounds we conclude that the Scriptures furnish no data for a chronological computation prior to the life of Abraham; and that the Mosaic records do not fix and were not intended to fix the precise date either of the Flood or of the creation of the world."

Hugh Ross, "Calibrating the Genesis 11 Genealogy." *Facts for Faith* Issue 10 (Q3 2002): p. 21.

Patricia Barnes-Svarney, editor, *The New York Public Library Scientific Desk Reference.* New York, NY: Macmillan, 1995.

Specific references on evolution, creationism and intelligent design are listed under the next three chapters.

Chapter 13: Evolving Origins

1. Adrian Melott, "Intelligent Design Is Creationism in a Cheap Tuxedo." *Physics Today* Volume 55, Issue 6 (2002). Also available on-line at www.physicstoday.org/vol-55/iss-6/p48a.html.

2. Lee Smolin, *Three Roads to Quantum Gravity* (New York, NY: Basic Books, 2001), pp. 197–199.

3. Gordon Kane, *Supersymmetry: Unveiling the Ultimate Laws of the Universe* (Cambridge, Massachusetts: Helix Books, 2000), pp. 142–147.

4. Carl Sagan, *Pale Blue Dot* (New York, NY: Random House, 1994), p. 38.

5. Phillip E. Johnson, *The Wedge of Truth* (Downers Grove, Illinois: InterVarsity Press, 2000), pp. 96–97.

Alan Hayward, *Creation and Evolution: Rethinking the Evidence from Science and the Bible*. Minneapolis, Minnesota: Bethany House, 1985, (1995 printing).

Bryan Sykes, *The Seven Daughters of Eve*. New York, NY: Norton, 2002. While Sykes seems to be a supporter of the naturalistic evolution paradigm, this is an excellent book detailing with the use of genetics to explore our ancestry. The problems related with mankind's late appearance and how we supposedly evolved from older primates are not addressed.

Fazale R. Rana, "Diseases Follow Human Origin and Expansion." *Connections* (V5, N2: Second Quarter 2003), pp. 1, 9. Also available on-line at www.reasons.org/resources/connections/2003v5n2/index.shtml?main.

Fazale R. Rana, "The Leap to Two Feet: The Sudden Appearance of Bipedalism." *Facts for Faith* Issue 7 (Q4 2002): pp. 32–41. Also available on-line at www.reasons.org/resources/fff/2001issue07/index.shtml?main#bipedalism.

Fazale R. Rana, "Repeatable Evolution or Repeated Creation?" *Facts for Faith* Issue 4 (Q4 2000): pp. 12–21. Also available on-line at www.reasons.org/resources/fff/2000issue4/index.shtml?main#repeatable_evolution.

Getting the Facts Straight: A Viewer's Guide to PBS's Evolution. Seattle, Washington: Discovery Institute, 2001. Also available on-line at www.ReviewEvolution.com.

While claiming to not speak to the religious realm, the PBS series repeatedly did so. As *Getting the Facts Straight* details, PBS' message seemed to be "Religion that fully accepts Darwinian evolution is good. Religion that doesn't is bad" (p. 14). The series also fails to "…present accurately and fairly the scientific problems with the evidence for Darwinian evolution" (p. 9) and it systematically omitted the "…disagreements among evolutionary biologists themselves…" (p. 9). To top this off this is a government subsidized channel: Your tax money funded a series that was designed to "co-opt existing local dialogue" (p. 15). In other words, the series was promoting a political agenda and served as evangelism for naturalistic evolution.

Hugh Ross, "Anthropic Principle: A Precise Plan for Humanity." *Facts for Faith* Issue 8 (Q1 2002): pp. 24–31. Also available on-line at www.reasons.org/resources/fff/2002issue08/index.shtml?main#a_precise_plan.

Hillary Mayell, "Documentary Redraws Humans' Family Tree." Article for *National Geographic News* on-line at news.nationalgeographic.com/news/2002/12/1212_021213_journeyofman.html, January 21, 2003.

Jennifer Viegas, "Study: Human DNA Neanderthal-Free." Article for *Discovery News* on-line at dsc.discovery.com/news/briefs/20030512/Neanderthal.html, May 12, 2003.

John N. Clayton, "Are Chimps 98.5% Human?" *Does God Exist?* Volume 28, Number 3 (May/June 2001). Also available on-line at www.doesgodexist.org/MayJun01/AreChimps98.5Human.html.

Jonathan Wells, *Icons of Evolution: Science or Myth? Why Much of What we Teach About Evolution is Wrong.* Washington, DC: Regnery, 2002.

Lee Spenter, *Not by Chance!* Brooklyn, NY: Judacia Press, 1998.

Michael J. Behe, *Darwin's Black Box.* New York, NY: Touchstone, 1996 (1998 printing).

Michael Denton, *Evolution: A Theory in Crisis.* Bethesda, Maryland: Adler & Adler, 1986 (1996 printing).

Neil Broom, *How Blind is the Watchmaker? Nature's Design & the Limits of Naturalistic Science.* Downers Grove, Illinois: InterVarsity Press, 2001.

Percival Davis and Dean H. Kenyon, *Of Pandas and People: The Central Question of Biological Origins.* Dallas, Texas: Haughton, 1993 (fourth printing 1999).

Phillip E. Johnson, *Darwin on Trial.* Downers Grove, Illinois: InterVarsity Press, 1993.

Phillip E. Johnson, *Defeating Darwinism by Opening Minds.* Downers Grove, Illinois: InterVarsity Press, 1997.

Phillip E. Johnson, *The Wedge of Truth.* Downers Grove, Illinois: InterVarsity Press, 2000.

Richard Deem, "Evolutionary Descent of Man Theory: Disproven by Molecular Biology." Detailed and well-documented paper at www.evidence.info/design/descent.html.

William A. Dembski, *Intelligent Design: The Bridge Between Science & Theology.* Downers Grove, Illinois: InterVarsity Press, 1999.

Chapter 14: The Genesis Question

1. Neil Broom, *How Blind is the Watchmaker?* (Downers Grove, Illinois: InterVarsity Press, 2001), pp. 72–73.

2. Ken Ham, *Creation Evangelism for the New Millennium* (Master Books, 1999), p. 127.

3. "Young-Earth Creationism: Interview with Dr. John Mark Reynolds." Creation Update radio program, Air-date: 3–7–2002. Archived on-line at www.reasons.org/resources/multimedia/rtbradio/cu_archives/index.shtml?main.

   It is interesting to note that in this interview and one with young-earth astronomer, Danny Faulkner (aired 9–26–2002, see same website) that they seem to agree that young-earth science is weak. Also, they seem willing to engage in reasonable and honest discussion on this subject rather than the emotional-charged, one-sided efforts that usually mark these debates. The only people who benefit from the common hostile discussions on this subject are naturalists and other opponents of Christians.

4. Goddard Space Flight Center, "New Image of Infant Universe Reveals Era of First Stars, Age of Cosmos, and More." Article posted at www.gsfc.nasa.gov/topstory/2003/0206mapresults.html on February 11, 2003. Article describes how NASA's Wilkinson Microwave Anisotropy Probe (WMAP) has shown the universe's age to be "…13.7 billion years old, with a remarkably small one percent margin of error."

5. "Arguments we think creationists should NOT use." Answers in Genesis website at www.answersingenesis.org/Home/Area/faq/dont_use.asp.

6. Andy Butcher, "He Sees God in the Stars." *Charisma* Volume 28, Number 11 (June 2003): pp. 38–44. On page 44, this article gives a good example of the dichotomy Christians often find themselves in:

    "Mark Clark, professor of political science at California State University…says [Dr. Hugh] Ross' ministry [Reasons to Believe] 'saved my faith.' He had embraced young earth creationism because 'it seemed to make sense.' But he found himself developing 'Christian schizophrenia' because he could not bring his weekend and workday worlds, and their conflicting realities, together."

7. "Which Came First?" *Answers Update* (June 2002), p.4.

8. Carl Wieland, "Warning to Families!" *Answers Update* (V5, N2: February 1998).

9. Ken Ham, "What's Wrong With 'Progressive Creation?'" *Answers Update*, (V4, N10: October 1997).

10. Kathy Ross, "Field Report." *Connections* (V3, N2: Second Quarter 2001), p. 3.

Ross is the wife of Dr. Hugh Ross, astronomer and founder of Reasons to Believe (www.reasons.org). Dr. Ross is a frequent target of young-earthers because he has so effectively shown the problems with their theory. When other intelligent design theorists begin to do the same, then what? As Chapter 15 describes, young-earthers already attack intelligent design because of its adherence to old-age, but not with as much vigor as they attack Dr. Ross. Sooner or later intelligent design theorists will have to remove this stumbling block (young-earthism), enhancing the problems intelligent design poses to its young-earth adherents (see Chapter 15). More on the Christian Research Institute can be found at www.equip.org.

11. See Note 3 above.

12. David G. Hagopian, editor, The Genesis Debate: *Three Views on the Days of Creation* (Mission Viejo, CA: Crux Press, 2000), p. 18.

Alan Hayward, *Creation and Evolution: Rethinking the Evidence from Science and the Bible*. Minneapolis, Minnesota: Bethany House, 1985, (1995 printing).

Hayward prefaces the creation-date issue by stating, "...I shall be obliged to oppose the notion that the Earth is young. But I shall not *attack* it; one does not attack one's own friends...I hope my allies will view me as exhorting them rather than attacking them...For recent-creationists *are* my friends and allies. Let there be no mistake about that. The things we have in common are much more important than those on which we differ. We share a belief in an inspired Bible. We agree that Darwin was mistaken, and that God is the Creator of every living thing. Compared with this, the question of the age of the Earth pales into insignificance."

Bill T. Arnold, *Encountering the Book of Genesis*. Grand Rapids, Michigan: Baker Books, 1998.

Arnold comments on old-earth creationism, "This view recognizes the problems of using biblical genealogies for precise chronology. In light of the overwhelming evidence in support of the antiquity of Earth, progressive creationists accept the traditional day-age interpretation of the creation days in Genesis...In this way [they] emphasize the complementarity between Genesis and modern science...all accept the old earth and some degree of microevolutionary theory...But [they] reject macroevolution and organic evolution because of a lack of scientific evidence."

Chris Stassen, "A Criticism of the ICR's Grand Canyon Dating Project." Website at www.talkorigins.org/faqs/icr-science.html.

Don Stoner, *A New Look at an Old Earth: Resolving the Conflict Between the Bible & Science*. Eugene, Oregon: Harvest House, 1997 edition. On-line version at answers.org/newlook/NEWLOOK.HTM#New.

Hill Roberts, "Evidences That Have Led Many Scientists to Accept An Ancient Date for Creation of the Earth and Universe." Articles, on-line book and other resources on the creation-date issue at lordibelieve.org/page15.html.

Hugh Ross, *Creation and Time: A Biblical and Scientific Perspective on the Creation-Date Controversy*. Colorado Springs, Colorado: NavPress, 1994. This is the most thorough book on this topic.

Hugh Ross, *The Genesis Question*, 2nd ed. Colorado Springs, Colorado: NavPress, 2001. Probably one of the best looks at Genesis-science issues and flood interpretations.

Kenneth Richard Samples, "Creedal Controversy: The Orthodoxy of Days." *Facts for Faith* Issue 7 (Q4 2001): p. 54. Also available on-line at www.reasons.org/resources/fff/2001issue07/index.shtml?main#creedal.

Kenneth Richard Samples, "Presbyterian Church in America Discusses Creation Days: Study Committee Affirms Diversity of Views." *Facts for Faith* Issue 3 (Q3 2000): pp. 62–63. The actual PCA study is on-line at www.reasons.org/resources/apologetics/pca_creation_study_committee_report.shtml?main.

Matthew S. Tiscareno, "Is There Really Scientific Evidence for a Young Earth?" Website at www.gps.caltech.edu/~tisco/yeclaimsbeta.html.

NASA Space Interferometry Mission Science, "Parallax." Website at sim.jpl.nasa.gov/science/parallax.html that details the upcoming space probe "scheduled for launch in 2009, [that] will determine the positions and distances of stars several hundred times more accurately than any previous program" and the science and mathematical concepts behind it.

Philip Plait, *Bad Astronomy*. New York, NY: Wiley, 2002. Plait spends a chapter addressing young-earth astronomy arguments.

Richard Deem, "Scientific Evidence for the Age of the Universe." Website at www.evidence.info/creation/ageofuniverse.html. See also by same author "Day-Age Genesis One Interpretation" at www.evidence.info/creation/day-age.html,

"Biblical Evidence for Long Creation Days" at www.evidence.info/creation/long-days.html and "Biblical Defense of Long Creation Days" at www.evidence.info/creation/dayagedefense.html.

Roger C. Wiens, "The Dynamics of Dating: The Reliability of Radiometric Dating Methods." *Facts for Faith* Issue 7 (Q4 2001): pp. 10–19. Also available on-line   at   www.reasons.org/resources/fff/2001issue7/index.shtml?main# dynamics_of_dating. A similar article by Wiens, "Radiometric Dating: A Christian Perspective," is on-line at www.asa3.org/ASA/resources/Wiens.html.

Roger C. Wiens and G. Brent Dalrymple, *The Age of the Earth.* Stanford, CA: Stanford University Press, 1991.

Ronald L. Numbers, *The Creationists: The Evolution of Scientific Creationism.* Berkley, California: University of California Press, 1992.

Samuel R. Conner and Hugh Ross, "The Unraveling of Starlight and Time." Final Revision March 22, 1999. On-line at www.reasons.org/resources/apologetics/unravelling.shtml?main. A clear refutation of Dr. D. Russell Humphreys' claim that he has discovered an alternative cosmology which claims to resolve the long-standing problem for the young-earth movement of "how light could travel billions of light years from distant galaxies during the passage of only a few thousand years of Earth time."

"Westminster Theological Seminary and the Days of Creation: A Brief Statement." *Facts for Faith* Issue 7 (Q4 2001): pp. 55–57. Also available on-line at www.wts.edu/news/creation.html.

Chapter 15: A Designed Universe?

1.   William A. Dembski, editor, *Mere Creation* (Downers Grove, Illinois: Inter-Varsity Press, 1998), p. 94.

2.   Robert Wright, "The 'New' Creationism." On-line at slate.msn.com/ ?id=104349, April 16, 2001. References 2–4 are good examples of design criticisms that do not address the issues and are full of fallacies.

3.   Adrian L. Melott, "Intelligent Design Is Creationism in a Cheap Tuxedo." On-line at www.physicstoday.org/vol-55/iss-6/p48a.html, June 2002.

4.   Mano Singham, "Philosophy Is Essential to the Intelligent Debate." On-line at www.physicstoday.org/vol-55/iss-6/p48b.html, June 2002.

5.   William A. Dembski, *Intelligent Design: The Bridge Between Science & Theology* (Downers Grove, Illinois: InterVarsity Press, 1999), pp. 122–152.

6.   Hugh Ross, "Summary of Reasons To Believe's Testable Creation Model." On-line at www.reasons. org/resources/apologetics/testablecreationsummary.shtml   and   also   see "Abbreviated Version of the New, Testable, Creation Model" at www.reasons.org/resources/multimedia/interview/19990919.ram (audio file).

7.   Charles Darwin, *Origin of Species*, 6th ed. (New York, NY: New York University Press, 1988), p. 51.

8.   John G. West, Jr. "Intelligent Design and Creationism Just Aren't the Same."   On-line   at   www.discovery.org/viewDB/index.php3?program=CRSC&command=view&id=1329, January 9, 2003.

West writes: "The two most prominent creationist groups, Answers in Genesis Ministries (AIG) and Institute for Creation Research (ICR) have criticized the intelligent design movement (IDM) because design theory, unlike creationism, does not seek to defend the Biblical account of creation. AIG specifically complained about IDM's 'refusal to identify the Designer with the Biblical God' and noted that 'philosophically and theologically the leading lights of the ID movement form an eclectic group.' Indeed, according to AIG, 'many prominent figures in the IDM reject or are hostile to Biblical creation, especially the notion of recent creation….' Likewise, ICR has criticized ID for not employing 'the Biblical method,' concluding that 'Design is not enough!' Creationist groups like AIG and ICR clearly understand that intelligent design is not the same thing as creationism."

Note here then when AIG and the ICR say "Biblical creation" and so forth they are referring to their young-earth theory. This begs the question of whether or not their theory is valid. They do rightly point out the problems with the IDM stopping short concerning who the designer is and allowing any beliefs in their "eclectic" group, but make no mistake, the age issue is their paramount objection. Young-earthers focus their attacks on those few people whom have demolished their position and they make it sound as if those few are in the minority. If (or more likely *when*) the rest of the IDM does address this, however, the young-earthers may find their theory marginalized to the point of obscurity.

Fuzale R. Rana, "FYI: I.D. in DNA—Deciphering Design in the Genetic Code." *Facts For Faith* Issue 8 (Q1 2002): pp. 14–23. Also available on-line at www.reasons.org/resources/fff/2002issue8/index.shtml?main#deciphering_design.

Guillermo Gonzalez, "The Measurability of the Universe: A Record of the Creator's Design." *Facts for Faith* Issue 4 (Q4 2000): pp. 42–48. Also available on-line at www.reasons.org/resources/fff/2000issue04/index.shtml?main#measure_universe.

Hugh Ross, "Anthropic Principle: A Precise Plan for Humanity." *Facts for Faith* Issue 8 (Q1 2002): pp. 24–31. Also available on-line at www.reasons.org/resources/fff/2002issue08/index.shtml?main#a_precise_plan.

Hugh Ross, *The Creator and the Cosmos*, 3rd ed. Rev. Colorado Springs, Colorado: NavPress, 2001.

Hugh Ross, "More Than Intelligent Design." *Facts for Faith* Issue 10 (Q3 2002): p. 64. Also available on-line at www.reasons.org/resources/fff/2002issue10/index.shtml?main#more_than_id.

Jimmy H. Davis and Harry L. Poe, *The Designer Universe: Intelligent Design and the Existence of God*. Nashville, Tennessee: Broadman and Hollman, 2002.

Michael J. Behe, *Darwin's Black Box*. New York, NY: Touchstone, 1996 (1998 printing).

Michael A. Corey, *The God Hypothesis: Discovering Design in Our Just Right Goldilocks Universe*. Rowman & Littlefield, 2002.

Percival Davis and Dean H. Kenyon, *Of Pandas and People: The Central Question of Biological Origins*. Dallas, Texas: Haughton, 1993 (fourth printing 1999).

Phillip E. Johnson, *Defeating Darwinism by Opening Minds*. Downers Grove, Illinois: InterVarsity Press, 1997.

Phillip E. Johnson, *The Wedge of Truth*. Downers Grove, Illinois: InterVarsity Press, 2000.

Ralph O. Muncaster, *Dismantling Evolution: Building the Case for Intelligent Design*. Eugene, Oregon: Harvest House, 2003.

William A. Dembski and James M. Kushiner, editors, *Signs of Intelligence: Understanding Intelligent Design*. Grand Rapids, Michigan: Brazos Press, 2001.

Some related resources can be found in Chapter 6 and 13.

Chapter 16: Connections

1.   J.R.R. Tolkien, *The Lord of the Rings: The Fellowship of the Ring* (New York, NY: Ballantine Books, 1965), p. 76.

Appendix A: Beyond Reality

André Kole and Jerry MacGregor, *Mind Games: Exposing Today's Psychics, Frauds, and False Spiritual Phenomena*. Eugene, Oregon: Harvest House. Debunks some aberrant teachings on demons.

Brian Greene, *The Elegant Universe*. New York, NY: Vintage Books, 1999.

Hank Hanegraff, *The Covering*. Nashville, Tennessee: W Publishing, 2002. Through Hanegraff's discussion of Ephesians 6:10–18, he discusses some of the aberrant beliefs surrounding demons and the supernatural.

Hugh Ross, *Beyond the Cosmos*, 2rd ed. Rev. Colorado Springs, Colorado: Nav-Press, 1999.

Hugh Ross, Kenneth Samples and Mark Clark, *Lights in the Sky and Little Green Men*. Colorado Springs, Colorado: Navpress, 2002.

Ken Grimes and Alison Boyle, "The Universe Takes Shape." *Astronomy* Volume 30, Number 10 (October 2002): pp. 34–39.

J. Richard Gott, *Time Travel in Einstein's Universe*. New York, NY: Houghton Mifflin, 2002. Gott pointed out the interesting example from H.G. Wells.

Steve Nadis, "When Branes Collide." *Astronomy* Volume 30, Number 5 (May 2002): pp. 34–39.

Appendix B: Science in the Bible

Hugh Ross, *The Creator and the Cosmos*, 3rd ed. Rev. Colorado Springs, Colorado: NavPress, 2001.

Hugh Ross, *The Genesis Question*, 2nd ed. Colorado Springs, Colorado: NavPress, 2001. Very extensive look at the connections between science and the first eleven chapters of Genesis.

Appendix C: Prehistory

Bryan Sykes, *The Seven Daughters of Eve*. New York, NY: Norton, 2002.

Chris Scarre, editor, *Past Worlds: Atlas of Archaeology*. HarperCollins, 2003.

Don Richardson, *Eternity in their Hearts*. Ventura, CA: Regal Books, 1983. An interesting look at how Christian beliefs are found in ancient peoples who never knew of Christianity. Suggests a common source of knowledge from early in man's history when they shared the same beliefs and the ability to find "fingerprints" of God around us and in us.

Fazale R. Rana, "Diseases Follow Human Origin and Expansion." *Connections* (V5, N2: Second Quarter 2003), pp. 1, 9. Also available on-line at www.reasons.org/resources/connections/2003v5n2/index.shtml?main.

Hillary Mayell, "Documentary Redraws Humans' Family Tree." Article for *National Geographic News* on-line at news.nationalgeographic.com/news/2002/12/1212_021213_journeyofman.html, January 21, 2003.

Hugh Ross, "Calibrating the Genesis 11 Genealogy." *Facts for Faith* Issue 10 (Q3 2002): p. 21. See also the notes on genealogies for Chapter 12.

Hugh Ross, *The Genesis Question*, 2nd ed. Colorado Springs, Colorado: NavPress, 2001.

James Lawrence Powell, *Night Comes to the Cretaceous*. New York, NY: Harcourt Brace,1998.

J.M Adovasio and Jake Page, *The First Americans: In Pursuit of Archaeology's Greatest Mystery*. New York, NY: Random House, 2002.

Richard Deem, "Evolutionary Descent of Man Theory: Disproven by Molecular Biology." Detailed and well-documented paper at www.evidence.info/design/descent.html.

Geoffrey Barraclough, editor, *Atlas of World History*. HarperCollins, 2001.

Appendix D: More on Young-Earth Creationism

1. For information on the highly suspect dinosaur artifacts and sightings such as the "Ica Stones," "Acambaro Figurines" and "Sightings of Pterodactyls" see "An Open Letter about Kent Hovind's Seminar" at bibleandscience.com/hovind.htm, "Ica Stones: Yabba-Dabba-Do!" at www.csicop.org/si/2002–09/strange-world.html and "Ica stones" at skepdic.com/icastones.html.

2. Hugh Ross, *Creation and Time: A Biblical and Scientific Perspective on the Creation-Date Controversy* (Colorado Springs, Colorado: NavPress, 1994), pp. 114–115.

See the Notes for Chapter 14 for references and resources on young-earth creationism.

Appendix E: Where did God Come from Anyway?

Hugh Ross, *Beyond the Cosmos*, 2nd ed. Rev. Colorado Springs, Colorado: NavPress, 1999.

Kenneth D. Boa and Robert M. Bowman, Jr., *20 Compelling Evidences That God Exists*. Tulsa, Oklahoma: River Oak, 2002.

Peter Kreeft & Ronald K. Tacelli, *Handbook of Christian Apologetics*. Downers Grove, Illinois: InterVaristy Press, 1994.

Appendix F: The Case for Christ

1. Michael Grant, *Jesus: An Historian's Review of the Gospels* (New York, NY: Charles Schribner's Sons, 1977), p. 176.

2. Carl Braaten, *History and Hermeneutics*, vol. 2 of *New Direction in Theology Today*, editor William Hordern (Philadelphia, PA: Westminster Press, 1966), p. 78.

3.   Michael Green, *The Empty Cross of Jesus* (Downers Grove, Illinois: InterVarsity Press, 1984), p. 97.

Gary R. Habermas, *The Historical Jesus: Ancient Evidence for the Life of Christ.* Joplin, Missouri: College Press, 1996.

Hank Hanegraff, *The Third Day.* Nashville, Tennessee: W Publishing, 2003.

J.N.D. Anderson, *The Evidence for the Resurrection.* Downers Grove, Illinois: InterVarsity Press, 1966.

Kenneth D. Boa and Robert M. Bowman, Jr., *20 Compelling Evidences That God Exists.* Tulsa, Oklahoma: River Oak, 2002.

Lee Strobel, *The Case for Christ.* Grand Rapids, Michigan: Zondervan, 1998.

Peter Kreeft & Ronald K. Tacelli, *Handbook of Christian Apologetics.* Downers Grove, Illinois: InterVaristy Press, 1994.

About the Author

Darrick Dean is a graduate of Geneva College (www.geneva.edu) where he earned is B.Sc. in Engineering (Concentration in Mechanical Engineering, Minor in Mathematics). He has had articles on space exploration published in *Ad Astra*, *Space Times* and *Spaceviews*, contributed to *Does God Exist?* and formerly penned the "Critical Thinking" column for the *New Castle News*.

An active supporter of space exploration, he spent 11 years as a member of the National Space Society (www.nss.org) and is listed as a contributor to NASA's 1989 Space Exploration Initiative Outreach program. In 1998, he was awarded the national Arthur L.Williston Award by the American Society of Mechanical Engineers (www.asme.org) professional organization for the paper "The Benefits and Necessity of Manned Exploration of Frontiers as Compared to Unmanned Efforts."

A former member of ASME, he is now a member of the American Society of Quality (www.asq.org) professional organization and works in industry in the quality field. He maintains a website dedicated to many of the topics covered in this book at www.geocities.com/darrickdean.

0-595-29185-6

Printed in the United States
37308LVS00004B/92

9 780595 291854